Grassroots Responses
to Extractivism

Grassroots Responses
to Extractivism

Case Studies from Around the World

Felix Mantz, Mariko Frame, and Samuel Grant

BLOOMSBURY ACADEMIC
LONDON • NEW YORK • OXFORD • NEW DELHI • SYDNEY

BLOOMSBURY ACADEMIC
Bloomsbury Publishing Plc
50 Bedford Square, London, WC1B 3DP, UK
1385 Broadway, New York, NY 10018, USA
29 Earlsfort Terrace, Dublin 2, Ireland

BLOOMSBURY, BLOOMSBURY ACADEMIC and the Diana logo are trademarks of
Bloomsbury Publishing Plc

First published in Great Britain 2025

Cover design by Adriana Brioso
Cover image: *Spring*, 1977 (oil on canvas) by Galambos, Tamas. All Rights Reserved, 2024.
Bridgeman Images.

ISBN: HB: 978-1-3503-3160-0
PB: 978-1-3503-3164-8
ePDF: 978-1-3503-3162-4
eBook: 978-1-3503-3161-7

Typeset by Deanta Global Publishing Services, Chennai, India
Printed and bound in Great Britain

To find out more about our authors and books visit www.bloomsbury.com
and sign up for our newsletters.

Contents

Illustrations

Figures

Tables

I

Introduction

Framing Socio-ecological Crises
in the World-System

Living Histories and Deep Structures

Mariko Frame and Felix Mantz

From local activism against land grabbing in Liberia and deforestation in Cambodia to Indigenous alternatives of territorial sovereignty, this anthology contains narratives of frontline resistances to extractivism and the escalating socio-ecological crises of the twenty-first century. It showcases movements and perspectives from the "Global South," spanning Africa, Latin America, and Asia, as well as Indigenous and Black voices in the "Global North." Through place-based case studies, this volume brings together diverse responses to the multiple forms of extractivism and socio-ecological crises that are simultaneously distinct but, at a deeper level, arise from trends both systemic and global in nature. In this volume, the themes of ecological destruction and alienation, social erosion, and political conflict emerge and overlap in case study after case study, highlighting how socio-ecological crises are deeply rooted in societies and economies across the globe organized under the dominant mode of production, industrial capitalism. These case studies also illustrate the highly stratified nature of capitalism, where modernization, wealth, and development for the few results in impoverishment, and ecological and social entropy for others, in particular for the urban and rural poor, Indigenous peoples, and along gendered and racial lines.

A mixture of successes and failures, the stories contained in this volume demonstrate the power of grassroots resistances, as well as the creativeness and variety of strategies and perspectives. As such, the case studies in this volume underscore the "pluriversal" nature of grassroots movements globally. Rather than a homogenous "multitude" of grassroots movements, the case studies demonstrate how resistances and alternatives are constituted by many kinds of worlds, ontologies, and ways of being, a heterogenous reality that some have come to characterize as "pluriversal." Creating a world in which many worlds fit, as the Zapatistas would say, thus emerges as a central political project when we consider these cases together and in relation to one another. Further, pluriverse, drawing from Indigenous relational worldviews, implies the existence of many interconnected worlds, such as the interdependency and coexistence of the human world with the natural world and spiritual world (Querejazu, 2016). In many of the case studies, recognition of this interconnectivity, both spiritually and pragmatically, plays a key role in shaping the resistances and alternatives, including in their deeper critique against homogenizing forces of modernity and globalization.

While pluriversal resistances and movements for alternatives across the planet face specific circumstances to an extent, their challenges also arise in response to problems that are world-systemic and often fundamentally similar in nature. The crises examined in this volume range from land grabbing and deforestation to forest conservation, mineral extraction, hydroelectric dam projects, and beyond. We term such crises socio-ecological because they involve contestation, struggle, and conflict that are fundamentally about how to organize the relationship between humans and the ecologies of which they are part. Key world-systemic characteristics underpinning these crises include (1) capitalism as the dominant mode of production; (2) colonial, racialized, and patriarchal systems of power and knowledge; (3) the continued hegemonic control of imperial countries over technology, finance, military, and ideology; (4) orthodox and Eurocentric conceptions of development, progress, and modernization; (5) the biophysical basis of industrial capitalism—which is best described as extractivism; (6) the political economy of neoliberal globalization and the power of transnational capital; (7) the hegemony of the nation-state; and finally (8) domestic capital and complex class and caste dynamics within the Global South. These key characteristics, in some shape or form, underpin the multiple socio-ecological crises spanning the globe, shaping dynamics of imperialism, stratification, and inequality.

The following sections provide a broad overview of these dynamics, histories, and structures with the goal of framing the heterodox contributions featured in this volume. This framing begins with the deep structures of coloniality and modernity; the relations of race, gender, and capital they engender; and how they enable complex forms of erasure of people, ecologies, and knowledges. Based on this discussion of coloniality and its deep structural formations, the chapter then focuses on the dynamic relationships of more recent characteristics of the world-system, namely capitalism, developmentalism, neoliberal globalization, and the state.

Colonaility, Extractivism, and the Deep Roots of Socio-Ecological Crises

Many contributions to this volume place alternatives and resistance to socio-ecological crises in the context of long-standing struggles against colonialism and racism in their many forms. Anticolonial, decolonial, and Indigenous critiques are highly diverse and complex, emphasizing different dimensions and experiences of colonialism, racism, and imperialism. As a result, there are many concepts and theorizations of colonialism and its contemporary manifestations. This section synthesizes some of them through the concept of coloniality and its critique of modernity. While this should help readers contextualize the many contributions in this volume, not all authors featured here use the same vocabulary.

Coloniality and the Origin of the Modern International Order

Anticolonial, decolonial, and Indigenous perspectives often identify 1492 as a key origin point for the making of the modern international order (Quijano, 2000, 2007; Wynter, 2003; Lugones, 2007; Grosfoguel, 2013; Vázquez, 2017; Mignolo and Walsh, 2018; Alimonda, 2019; Francis, 2019). In 1492, a host of world-historical processes took place. First, Columbus arrived in Abya Yala (now the "Americas"), inaugurating a systematic process of genocide, conquest, and colonization that reduced the Indigenous population of Abya Yala from approximately 100 to 60 million to only about 6 million by 1650. This made way for European settlement, resource extraction, and industries (Grosfoguel, 2013; Figueroa Helland, Pérez Aguilera, and Mantz, 2021). This genocide was so comprehensive that it left a permanent mark in the Earth's geological history. Around 1610, there was a sudden drop in the Earth's atmospheric CO_2 levels, followed by a rapid and steady increase that has held on until today (Vázquez, 2017). This drop was most likely due to the genocidal depopulation of Abya Yala which initially allowed carbon-absorbing forests to regrow where subsistence agriculture and Indigenous cultivation took place before settler expansion, mineral extraction, and monocrop agriculture destroyed these forests and landscapes, altering the climate at an unprecedented scale (Vázquez, 2017). This genocidal and ecocidal process catalyzed the enslavement of tens of millions of Africans who were kidnapped from their homes and transported via the Middle Passage to the colonies in the Americas. There, they were put to work primarily as cheap and disposable labor on expanding plantation economies whose produce and profits would feed the industrialization of Europe (Grosfoguel,

2013; Rodney, 2018). Within Europe in 1492, a radical social reordering took place. On the Iberian Peninsula, a war of extermination was waged against Muslims and Jews (Grosfoguel, 2013). Across Europe more broadly (and then elsewhere), a second war of extermination was waged against women, commonly known as the "witch hunts." The witch hunts erased non-Christian spiritualities, knowledges, and relations to ecologies; broke the relative power and autonomy of rural dwellers to facilitate the enclosure of commonly managed lands; and gave rise to modern heteropatriarchy (Federici, 2004; Grosfoguel, 2013).

It was thus in 1492, through force, conquest, and bloodshed, that the modern international order as we know it today was brought into life. This modern international order, as anticolonial, decolonial, and Indigenous perspectives highlight, did not start in the factories of Birmingham or Manchester but in the Atlantic with the conquest and colonization of Abya Yala. As Frantz Fanon (2011: 53–8) describes,

> This European opulence . . . was built on the backs of slaves, it fed on the blood of slaves, and owes its very existence to the soil and subsoil of the underdeveloped world. Europe's well-being and progress were built with the sweat and corpses of blacks, Arabs, Indians, and Asians. . . . Europe has been bloated out of all proportions by the gold and raw materials from such colonial countries as Latin America, China, and Africa. . . . Europe is literally the creation of the Third World. The riches which are choking it are those plundered from the underdeveloped peoples.

Where did the cotton, the silk, the timber, and the rubber that fed the machines of Europe come from if not the colonies? How were the profits which were mobilized to industrialize Europe accumulated if not through the enslavement, commodification, and trade of humans on account of their Blackness? How did European capitalists acquire knowledge on dying and weaving if not through the colonization of India and the subsequent destruction of its superior textile industry? How were the workers of Europe's factories fed and supplied if not through the sugar, tea, coffee, and tobacco from the colonies and the reproductive labor of domesticated women who were now working "triple shifts"—as wives, as mothers, and as workers in "productive industries" (Federici, 2004; Fanon, 2011; Rodney, 2018)?

The modern international order is only possible because of the establishment of European colonial dominance since 1492, a process that is ongoing today. As the testimonies and accounts in this volume make clear, this order can only be sustained by continuously reproducing its systems of power through erasure, extraction, and exploitation (Quijano, 2007; Grosfoguel, 2013). This is what we mean with "coloniality" (Quijano, 2000; Mignolo and Walsh, 2018). Crucially, what distinguishes *colonialism* from *coloniality* is that the latter does not require the presence of a formal colonial administration. Also, in contrast to *colonization*, coloniality is not a time period that has a start and end date. Rather it is a global system of power. When former colonies finally gained independence, they did not enter a neutral international order devoid of power relations. Instead, this order was shaped by and continued to operate on a complex web of economic, racial, cultural, gendered, epistemic, and anthropocentric hierarchies that had been developed since 1492. It was this order that entrapped, contained, and metabolized anticolonial movements for liberation. At the same time, we must recognize that not all colonies underwent decolonization. Settler colonies (often with their own empires), where the colonizers never left, continue to exist and even expand, including the United States, Canada, Australia, Israel, and New Zealand. The continuation of a modern international order through coloniality and settler colonialism means that we have never arrived in a "postcolonial" world (Figueroa Helland and Lindgren, 2016).

Logics of Coloniality

The modern/colonial international order operates on a system of classification that spans the Earth and her earthlings (Quijano, 2007). This system of classification enables some to assume power and control over others who can, through various ways, tools, and strategies, be subjugated, oppressed, and exploited.

Central to coloniality are three key logics: race, gender, and capital (Quijano, 2000; Lugones, 2007; Mignolo and Walsh, 2018). *Race*, as a system of classification, seeks to establish a hierarchical order of peoples and ecologies which determines degrees of Humanity based on a set of characteristics. These characteristics have changed historically but always envisioned European bourgeois men as the full Human, while women, non-whites, and ecologies are placed lower on this ladder of Humanity (Wynter, 2003; Mignolo and Walsh, 2018). Systems of race and racialization include, but also go beyond, phenotypes such as color and pigmentation. They are also based historically on shifting and overlapping ideas of who counts as civilized or savage, Christian or heathen, developed or undeveloped, modern or traditional, and advanced or backward (Wynter, 2003; Mignolo and Walsh, 2018). These logics identified any deviation from the full Human, the European man, as a sign of lack—lacking humanity, lacking legitimate knowledge, lacking civilization/civility, lacking rights, lacking development. The imposition of specific ideas of *gender* was entangled with race and took multiple forms (Lugones, 2007). For one, this could entail the invention and imposition of gender in the first place in societies where reproductive organs were not a basis for social organization and where binary understandings of men/women are alien. Second, it could mean the reordering of existing gender relations for the needs of the colonizers which often resulted in the erosion of women's power in noncolonial societies as well as the imposition of sexual binaries and heterosexuality at the expense of fluid sexualities (Lugones, 2007). Entangled with race and gender is the logic of *capital*. During his first voyage to Abya Yala, Columbus writes in his journal on October 19, 1492: "I believe that there are many herbs and many trees [in the 'New World'] that are worth much in Europe for dyes and for medicines; but I do not know them, and this causes me great sorrow" (Columbus, 2003: 123). From the beginning, colonialism was concerned with the generation of profit which requires the pricing, commodification, dismembering, and extraction of ecologies and people. Colonialism was motivated not only by the "civilizing" and Christianizing mission but also by the search for so-called "natural resources" and fertile lands for an emerging capitalist order (Alimonda, 2019; Francis, 2019). As discussed in more depth in the subsequent sections of this chapter, this process continues until today, as virtually all of the chapters in this volume highlight.

Underneath the logics of race, gender, and capital, which are central to the modern international order, is a knowledge system that is based on a binary worldview. This binary worldview envisions the world to be shaped by several oppositions whose two parts relate to each other hierarchically. These include but are not limited to Civilized/Savage, Christian/Heathen, Secular/Spiritual, Male/Female, Rational/Emotional, Modern/Traditional, Developed/Backward, Mind/Body, and Human/Nature (Moore, 2017; Mignolo and Walsh, 2018). For example, the Human/Nature binary sees the Human as the superior entity which properly relates to nature through invasion, domination, and extraction. Through the Human/Nature binary, ecologies were transformed into objects— the so-called "natural resources"—that only hold value when measurable through mathematical and scientific logics, when they can be quantified, partitioned, controlled, and made productive for Human use (Moore, 2017). The implications of these processes are outlined in detail, among others, in Chapter 2. In this chapter, Sherpa points to the deep rift between complex, spiritual, and non-dualistic relationships of Buddhist and Indigenous communities in the Himalayas on the one hand, and the developmentalist agenda of Indian central and regional state institutions that "clearly isolated nature from local cultures," on the other. Similarly, in Chapter 12, Pérez Aguilera highlights the clash between a Eurocentric and anthropocentric worldview and the complex multispecies relationalities of the Wixarika people of present-day Mexico. Pérez Aguilera further explains how Eurocentric and anthropocentric worldviews that seek to order and subjugate water bodies in the Basin of Mexico for modern developmentalist projects ultimately fail at this project of mastery over nature.

It is important to recognize that this category of the Human has never included all human beings, but only a tiny subset. Most humans—non-whites, women, non-Christians, the poor—were historically relegated to a sub-human status through the logics of race, gender, and capital. This sub-human status brought them closer to Nature according to the Western binary worldview (Moore, 2017; Mignolo and Walsh, 2018). We see this, for instance, in the description of the colonized as animals or animal-like, as beasts and savages, as part of flora and fauna, or as unpredictable, wild, and irrational, lacking the capacity of Reason—qualities that have also been inscribed in modern understandings of womanhood and femininity and which are used to legitimize colonial, capitalist, patriarchal, developmentalist, and extractive interventions.

This expulsion of the non-European from humanity manifests in various ways to legitimize the conquest of territories and ecologies across the globe. Crucially, this entailed declaring lands in Asia, Africa, and the Americas to be empty— "terra nullius" (Francis, 2019). Indigenous peoples' presence on these lands was seen as improper or insufficient. In many cases, they were conceptualized to be part of the landscape in the same way as nonhuman animals would be. Alternatively, colonizers claimed that the lands lacked proper land use, meaning that non-Europeans were not engaging with the land and its ecologies in the same way Europeans did: through separation, domination, and extraction. For example, mobile lifeways of pastoralists or the nurturing of polycultural food systems were often not recognized as legitimate and associated with an inability to properly subdue Nature. The idea of empty lands also referred (and continues to refer) to the absence of land ownership based on ideas of state, private, and individual property. This is especially important since communal land tenure systems were prevalent across the globe before European colonialism (Francis, 2019). This continues to be a key site of struggle today when state and corporate actors seek to grab lands and resources based on the claim that land is allegedly not owned and not used, ignoring communal forms of land tenure and existing forms of land use. This denial and erasure of existing relationships, systems of use, and forms of tenure regarding land legitimizes colonial land grabs, conquest, and extractivism (Figueroa Helland, Aguilera, and Mantz, 2021).

In addition, the process of colonial domination of ecologies was fundamentally a gendered process. As (eco)feminist scholars point out, depictions of women and nonhuman ecologies that would legitimize their domination were remarkably similar—both were associated with characteristics of irrationality, emotion, unpredictability, hypersexuality, and wildness and thus as needing to be tamed and controlled by white "rational" men (Plumwood, 1993). Francis Bacon, regarded commonly as the father of modern science, is a prime example. He promoted a patriarchal vision of subduing and conquering a feminized Nature which was to be "penetrated" and "enslaved" so that her "secrets" may be revealed by the (male) scientist (Federici, 2004).

Modernity's Erasure: Ecocide, Genocide, and Epistemicide

Part of the rationale for this volume is the alarming observation that throughout its 500-year history, the international order continues to attempt to erase what it perceives as non-modern. Again, this erasure works through the logics of race, gender, and capital and mobilizes a binary worldview to identify what is "non-modern" or "Other" (Vázquez, 2017; Dunlap, 2021). This Other—whether non-Western knowledges, non-extractive economic systems, nonhierarchical political structures, communal land relations, non-Christian spiritualities, non-European languages, sciences, and cosmologies, or other socio-ecological arrangements (e.g., non-commodified food systems, undammed rivers, or unexploited minerals), among others—is targeted for erasure and extraction through multiple mechanisms and for various reasons. Erasure not only seeks to disappear the Other but might also attempt to transform, consume, and incorporate it into the modern international order (Wolfe, 2006). For instance, as described by Nangiria and Lunch in Chapter 4, Mantz in Chapter 3, and Kaba and Sillah in Chapter 5, much of the land grabbing in Tanzania and Liberia that seeks to remove people from landscapes takes place under the guise of "conservation." Such land grabbing seeks to preserve a fantasy of African wilderness based on false Eurocentric understandings of the relationship between ecosystems and (African) humans who have maintained the ecological integrity, biodiversity, and aesthetic beauty of African landscapes.

Coloniality's erasure can be conceptualized through three interconnected tendencies which are key to understanding the planetary socio-ecological crises today: genocide, ecocide, and epistemicide (Grosfoguel, 2013; Dunlap, 2021; Figueroa Helland, Aguilera, and Mantz, 2021). While *genocide* refers to the deliberate and systematic elimination of a specific group of people, *ecocide* is the deliberate and systematic elimination of ecologies, landscapes, and ecosystems. Nou's testimony on the clearing of the Prey Lang Forest in Cambodia and its replacement by timber trees under the guise of "development" is an example of ecocide (Chapter 7). Globally, ecocide includes, among others, "the loss of 'thirteen million hectares of forests' every year, according to the United Nations, . . . [and] the 'desertification of 3.6 billion hectares' of dry lands" (Dunlap, 2021: 214). Ecocide is also expressed in "a recent study by the Intergovernmental Science-Policy Platform on Biodiversity and Ecosystem Services (IPBES), [which] concludes 'that around 1 million animal and plant species are now threatened with extinction, many within decades, more than ever before in human history'" (Dunlap, 2021: 214). This has culminated in what has been identified as the Sixth Mass Extinction in the Earth's history (Dunlap, 2021).

Epistemicide is the deliberate and systematic erasure of specific knowledges ("episteme" in Greek) and associated languages, spiritualities, cosmologies, and cultures on which political, economic, and social systems are based (Grosfoguel, 2013; Dunlap, 2021). For instance, epistemicide is coupled with spiritual invasion through Christianization, which together with the erasure of languages constitute what the anticolonial Gikuyu philosopher Ngũgĩ wa Thiong'o (2011) calls the "cultural bomb" of imperialism. Today, we see epistemicide, for instance, in the perpetual erasure of languages, a proxy for cultural and epistemic diversity. Figueroa Helland and his colleagues (2021: 27) explain that "[c]olonizing and 'modernizing' pressures . . . contribute to projections that 50–90% of the world's languages will be erased by 2100." Most of these are non-Western languages attuned to specific socio-ecologies and carry centuries of knowledge. Epistemicide can also mean the transformation or abstraction of specific knowledge systems. For example, in Chapter 11, Kassem finds that some forms of Islamic thought, including those that suggest non-anthropocentric relations, are often identified as "unscientific," "unreasonable," or "irrational" and thus become muted or erased. For Kassem, this shows how "modern frames of thinking and discourses of being filter into Muslim subjectivity and then into religious knowledge, discourses, and practices" (Chapter 11).

Throughout the 500 years of coloniality, we see modernity's tendency toward genocide, ecocide, and epistemicide, an impulse inherent to the modern international order. These three processes often occur simultaneously and can catalyze one another. For example, Israel's genocidal assault on Gaza after October 7, 2023, has entailed the destruction of every university in Gaza, as well as the systematic targeting of journalists and cultural figures as a way to erase Palestinian history, identity, and culture (epistemicide). At the same time, the genocide has catalyzed the destruction of Gaza's ecologies via the chemical pollution of water, soil, and air, and the destruction of farms, greenhouses, orchards, and groves, which undermines the life-sustaining capabilities of the land, producing unlivable conditions for Palestinians which further deepens disposession. As the chapters in this volume show, these entangled processes of erasure are actively confronted and resisted. For instance, the complex genocide-ecocide-epistemicide relationship constitutes the backdrop for the project of nourishing communal territories of life based on (Afro-)Indigenous resurgence in Abya Yala as laid out by Figueroa Helland, Martinez, Mendez, Rodríguez, Negrete, Correa, Gualinga, and Yvaire in Chapter 8.

The Political Economy of the Modern International Order

Capitalism and Socio-Ecological Crises across the Globe

Socio-ecological crises across the planet are rooted in the capitalist mode of production, a system that grows out of the 500-year history of the modern/colonial international order. Capitalism has several distinct defining features. First, it is a system where the means of production (e.g., factories, machines, financial capital, and land) are owned privately, sold in markets, and concentrated within the hands of a small portion of the population, the capitalists. Everyone else must sell their labor power for wages in order to survive. The second defining characteristic of capitalism is that the logic of production is fundamentally dominated by the ceaseless imperative of capitalists to accumulate more capital, or profit. Both features put capitalism directly at odds with socio-ecological sustainability. The first characteristic—the ownership of the means of production within the hands of certain people and consequently their capture of profits—shelters the privileged (usually along racial and gendered lines) from the environmental catastrophes that burden the poor and marginalized. The second feature—the ceaseless pursuit of a quantitatively limitless amount of profit—pushes capitalists to be continuously engaged in a process to appropriate and commodify labor and Nature as inputs for production. This is the essential reason why capitalism, since its beginning, is an expansionary system. Capitalism must constantly expand, seeking new markets as outlets for its goods, finding more "natural resources" for production purposes, searching for cheaper sources of labor, and many other reasons. This fundamental, expansionary, and extractive drive of capitalism, subsuming all socio-ecological systems in its path, is for obvious reasons highly problematic for the maintenance of global climate, environmental and hydraulic cycles that enable diverse communities of lifeforms, including humans and nonhumans, to flourish.

Extractivism as the Underside of Industrialization and Orthodox Development

Through colonialism, the concepts of progress, modernization, and civilization were unabashedly defined by European ideas of racial superiority and the domination of non-European societies (Figueroa Helland and Lindgren, 2016; Mignolo and Walsh, 2018). With the end of formal colonialism in many (but not all) parts of the world in the twentieth century, these concepts coalesced into a new definition of progress: *development*. Development retained Western-hegemonic ideals through emphases on supposedly linear, stagist pathways toward modernization and industrialization. With the continued power of First World capitalist nations in the post–Second World War era—particularly the United States—economic growth and rising consumption through market mechanisms became key ingredients in the new concept of development. Growing a country's gross national product (GNP) has become the primary economic goal for almost all countries globally. Further, underlying Western concepts of development was the belief that all developing countries can, and should, modernize along the same lines as the capitalist First World countries: particularly the United States. Former colonies were encouraged to actively emulate the Western model of free enterprise, with the contention that supposedly "traditional" societies would move through successive stages of modernization, economic growth, and technological advancement until they finally "caught-up" with the advanced industrial capitalist nations (McMichael, 2017).

The fixation of development strategies on economic growth through market mechanisms has, in direct contrast to the unbridled optimism of free market proponents, been devastating for ecologies globally. Against persistent promises of orthodox economists and techno-futurists, economic, industrial, and technological growth has not synergized in a virtuous cycle to reverse global ecological crises. Instead, global ecological crises are deepening. Significantly, hundreds of studies show that economic growth for countries everywhere has not decoupled from resource use or environmental impact in any sustained or significant way (Parrique et al., 2019). In contrast, the post–Second World War era has witnessed a drastic increase in the use of the Earth's resources, a period environmental scientists dub "The Great Acceleration" (IGBP, n.d.). As countries grow economically and industrialize, their environmental impacts do not decrease but escalate (Steffen et al., 2010; UNCTAD, 2012; Krausman et al., 2008). Perhaps most significantly, what has become evident is that the economic development and consumption levels of the Global North have been and continue to be largely built upon the destruction of ecologies globally (Frey et al., 2018; Dorninger et al., 2021). There is also increasing evidence which shows that as countries in the Global South industrialize, they too offshore the externalities of their industrial development and increased consumption elsewhere (Yu et al., 2014; Shandra et al., 2019)

In general, the promises of what were originally Eurocentric ideals of development in terms of endless growth, industrialization, increased consumption, and capitalism have now been wholeheartedly embraced by countries outside of the Western world, from more free market economies to economies with higher levels of state interventionism that can arguably be defined as state capitalist (Economics in Context Initiative, 2021). Investments from emerging economies in Asia, Latin America, and the "Middle East" compete with traditional patterns of investments from the United States and Europe in natural resource extraction, land, and infrastructure throughout the Global South. For example, as discussed in Chapter 6 and Chapter 7, land grabbing that devastates ecologies and communities in Cambodia is driven primarily by East and Southeast Asian capital. Further, one of the main tourist companies grabbing land for conservation in Loliondo, Tanzania, is a UAE-based company (Chapter 3 and Chapter 4). In the global flurry of investments, a wide variety of actors ranging from private capital to state capital compete for land and resources, resulting in what Klare (2012) terms "the race for what's left." Yet planetary socio-ecological crises are the irrevocable shadows of development ultimately built upon industrial capitalism.

Socio-ecologically degrading extractivism is the biophysical basis of capitalism, an empirical, concrete, physical reality that development ideals cannot mask in the twenty-first century, where ecological crises are escalating quickly toward irreversible tipping points of climate change, biodiversity loss, and other Earth-systemic transformations. In country after country, as cases from this volume discuss, development projects lead to the socio-ecological crisis nexus of environmental devastation, large-scale dispossession, conflict, and even genocide, where it is the poor and marginalized who pay the costs. For example, in Chapter 1, Raghu explains how foreign and domestic capital strategize to seize land of Indigenous adivasi and oppressed communities in order to extract and exploit the mineral wealth of Jharkhand (India). Resistance and autonomous organizing by these communities is directed specifically against capitalist extractivism and ecological imperialism pushed by regional, national, and transnational state and corporate actors. In Chapter 6, Frame presents interviews from Cambodian women dispossessed of their land in Phnom Penh for a development project on Boeung Kak Lake. The chapter outlines how they have suffered the loss of livelihoods, political oppression and conflict, and psychological and physical harm.

Neoliberalism as a Key Feature of Contemporary World-Systemic Extractivism and Imperialism

As discussed above, since its colonial inception, capitalism has been built upon the wholesale transformation of ecologies into "natural resources" via extractivism and the exploitation of distant lands and peoples. Yet, in the past fifty years or so, there have been seminal developments in the characteristics and manifestation of global capitalism which help to explain the intensifying and accelerating socio-ecological crises. Of central, though not exclusive, importance has been the neoliberal turn of capitalism. Neoliberal capitalist development is a very specific, radical reorientation of societies to rewrite their economies and in the process transform their socio-ecologies along free market lines. The primary characteristic of neoliberalism, whether in the Global North or in the Global South, is the removal of barriers to capital. As a general process, this has meant the transformation of economies across the globe according to free market principles, though the level of state interventionism in the economy still varies considerably from country to country. Nonetheless, there are some key identifiable neoliberal transformations, particularly with regard to globalization, such as through trade, investment, and finance, that have had critical impacts on socio-ecological relations.

A response to economic crises across the globe that began in the 1970s, neoliberalism was largely exported to countries in the Global South when these countries fell into the debt crises of the 1980s. Many were forced to turn to the World Bank and International Monetary Fund (IMF) for loans. These loans came with strings attached, known as conditionalities, or "structural adjustment policies" (SAPs). SAPs required these countries to restructure their economies and development pathways along what many critics would soon call "neoliberalism," and free market or laissez-faire economics. This neoliberal turn followed and undermined the substantial challenge posed to the global colonial/ capitalist order by the Third World movement and various forms of anticolonial economic nationalism. In the Global South, neoliberalism was translated into this new vision of development through a set of economic policies known as the Washington Consensus. The Washington Consensus encouraged trade liberalization, devaluation of currencies to stimulate export-growth, foreign direct investment (FDI), privatization, and the securing of property rights, among other policies. Overall, such policies favored the development of an outward-oriented, export economy organized through global markets with minimal regulation. In practice, neoliberalism has meant the removal of barriers to both domestic and, crucially, *foreign* capital.

For certain countries in the Global South, particularly in Sub-Saharan Africa and parts of Latin America, neoliberalism has meant the re-entrenchment of colonial patterns of production and export. Through SAPs, countries were required to specialize according to their "comparative advantage," which, according to the Western-dominated institutions of the World Bank and IMF, once again meant specializing in the export of natural resources. In addition to specialization, developing countries were pressured to pursue export-led growth through trade liberalization, removing any barriers to international trade such as tariffs. The World Bank and IMF argued that developing countries would experience economic growth that would also help them repay their debt. Consequently, trade liberalization and specialization appear to have deepened patterns of ecologically unequal exchange and the ecological degradation associated with it (Campbell and Loxley, 1989; Frame, 2023; Giljum and Eisenmenger, 2004). Even in emerging countries with more diversified economies, such as Malaysia, the export of resources from the extractivist sector is a critical part of their overall growth strategies.

Privatization is central to the neoliberal project and to ecological imperialism. Privatization entails placing land and resources that were previously held in common, or regulated by the government, into private hands, either domestic or, in places like Sub-Saharan Africa, foreign capital. The grabbing of land and resources throughout the Global South by foreign investors is no accident: neoliberal development policies strongly favor the rights of foreign investors under the argument that FDI increases the foreign exchange of low-income countries and results in economic growth (OECD, 2002). It is important to realize that beyond the foundations created by SAPs, neoliberalization is now an ongoing phenomenon promulgated by a wide variety of structures and policies. For example, multilateral trade agreements, disseminated through the WTO, and bilateral trade treaties encourage foreign investment by providing sweeping protections for investors, such as the investor-state dispute settlement mechanism

(ISDS). ISDS essentially allows investors, such as multinational corporations (MNCs), to sue governments if their policies have been deemed harmful to the investor. Companies in natural resource sectors (largely headquartered in the Global North) are most likely to initiate ISDS proceedings (against governments, largely in the Global South) (Perez-Rocha, 2021). Multilateral trade agreements have, of course, also been instigated and created at regional levels in the Global South, such as the ASEAN Free Trade Agreement of Southeast Asia, which pursues the same logic of trade and investment liberalization (ASEAN, n.d.).

Finally, while the IMF and World Bank have encouraged a host of domestic reforms in sectors, such as mining and land, that codify rights for foreign investors decades ago (UNCTAD, 2005), these same laws remain very much in place in many countries. The rewriting of laws and regulations regarding large-scale investment in land in order to attract foreign investors is fundamental to understanding a host of socio-ecological issues in the Global South, such as land grabbing, deforestation due to industrial agriculture, and pollution and waste associated with the outsourcing of manufacturing (Frame, 2023; Transnational Institute, 2013; Jorgenson, 2010, 2007, 2009; Grimes and Kentor, 2003). For example, local governments have allowed MNCs to take over resources, water, forests, land for mining operations, and even ownership of food by privatizing their genetic codes. Due to the acquisition of privatized state-owned enterprises (SOEs), major sectors such as mining and manufacturing have come under direct control of foreign investors, especially in Sub-Saharan Africa (Frame, 2023). Finally, financial liberalization—the removal of government regulations on their financial sectors—is a key feature of neoliberal capitalism. The financial capital of major banks, institutional investors such as hedge funds and pensions, and others, provides the necessary funding for the start-up and expansion of business activities, including funding fossil fuel expansion and deforestation (Forests and Finance, n.d.). As Chapter 10 by Brewer points out, we are in an era of imperial racial monopoly capitalism where finance is hegemonic and monopolies dominate the capitalist order, as profits are funneled to a small global elite.

The State

The contemporary nation-state form emerged in the context of European empire-building (Ndlovu-Gatsheni, 2015b; Bhambra, 2018; Nişancıoğlu, 2020). Further, while multiple forms of political organization existed across what is now the Global South, from egalitarian to highly centralized societies, the nation-state form that characterizes much of Africa, Latin America, and Asia today is the direct result of imperialism and colonialism (Ndlovu-Gatsheni, 2015a, 2015b; Rodney, 2018). The national boundaries and borders as well as the political, legal, and judicial apparatuses that were set up by colonial and imperial powers had the explicit goal of facilitating extraction, dividing Indigenous populations, maintaining order, and/or enabling white settlement. For example, the states that exist in Africa today are largely the outcome of the 1884/85 Berlin Conference, where European powers partitioned Africa into separate spheres of influence that were allocated to different colonial powers to colonize, settle, and exploit (Amin, 1994; Ndlovu-Gatsheni, 2015a, 2015b; Rodney, 2018). The purpose of states in the Global South was never to form independent, self-sustaining, or cohesive social, political, and economic units.

Following formal decolonization, in many places, the territorial units, and political and legal apparatuses established by the colonizers and imperialists were not changed (Fanon, 2011; Ndlovu-Gatsheni, 2015a, 2015b; Rodney, 2018). In fact, in many cases, the aim of decolonization was to take over these political structures—that is, replace the European governors with "Native" heads of state rather than abolish the state altogether and search for different forms of political organization. For instance, as Walter Rodney (2018: 268) explains, "the struggle to regain African independence was conditioned by the administrative framework of the given colonies" which included "adopting the boundaries carved by the imperialists." Ndlovu-Gatsheni (2015a: 52) adds that the state in Africa "remained operating like colonial states, unleashing violence on African people . . . [who] are still often treated like subjects rather than citizens by their leaders." While some changes to the territorial landscape did take place in Africa, none of these were "informed by any paradigm shift on borders, nationalism, nation, state, belonging, sovereignty and citizenship. European cognitive systems and models of ordering political spaces are still upheld" (Ndlovu-Gatsheni, 2018: 28). Nonetheless, the tremendous achievements of formal decolonization and difficulties of post-independence state-building in an openly hostile international order should not be brushed aside or trivialized.

Building on this colonial history, contemporary extractivism and imperialism continue to require the authoritarianism of the contemporary state. As David Harvey (2007) points out, because the roots of neoliberalism are the restitution of class power, neoliberalism is built upon institutions that are profoundly anti-democratic. For countries in the Global South, neoliberal developmental paths built upon industrial capitalism, liberalized foreign investment regimes, and export-led development that require the authoritarianism of the state to pave the way for extractivism by both domestic and foreign capital. As such, for many of the cases in this book, repression and resistance centers around the nexus between the state and foreign and domestic capital. While the state can at times reflect resistances from below or even anti-imperialist sentiments such as during the era of economic nationalism which followed the immediate period after formal decolonization, nearly all states throughout the world are integrated into a capitalist world-system and beholden to the forces of capitalist accumulation.

Place-Based Resistance: Introducing the Contributions

Existing Struggles

In Chapter 1, *Pratik Raghu* examines the Pathalgadi Movement in the Indian state of Jharkhand. Active between 2017 and 2019, this movement was primarily comprised of Indigenous adivasi communities who set up autonomous infrastructures in resistance to corporate and state extraction of minerals. Raghu shows how the Pathalgadi Movement challenged not only extractivism but also state sovereignty and Indigenous politics captured by dominant state, capital, and civil society processes. Specifically, he identifies (1) a dynamic nexus consisting of ecological imperialism, domestic capitalism, and national chauvinism; (2) class differences as an important obstacle for political unity within Indigenous struggles; and (3) the employment of a repertoire of subaltern ungovernability. This repertoire consists of strategies that both engage and challenge liberal democratic institutions in ways that constitute a form of non-/anti-liberal politics. In light of the recognition that movements against extractivism can exceed liberal democratic frames, this chapter calls on scholar-activists to reexamine their conventional practices, assumptions, and terms of solidarity with Indigenous and radical grassroots movements across the globe.

In Chapter 2, *Dawa Sherpa* studies contestations over water governance in the state of Sikkim, India. Located in the Himalayas, Sikkim has witnessed corporate and state capture of rivers through hydroelectric projects (HEPs) such as large dams. This has provoked widespread community organizing and resistance which Sherpa examines in this chapter. With great detail, the chapter centers the question of cosmology and the role it plays in resistance efforts against HEPs. Sherpa explores the complex entanglements of Indigenous Himalayan cosmology and Buddhist spirituality. Sharing similar forms of relationality to water, Indigenous and Buddhist protocols, institutions, and relations are revitalized in the context of organizing and nurturing opposition to HEPs. These seem to be especially strong foundations for such opposition due to the stark contrast between Buddhist and Indigenous Himalayan spiritualities and relationalities on the one hand, and state and corporate perceptions of water and approaches to development on the other. The chapter highlights not only the ways in which non-Western cosmologies translate into practices of resistance but also how they can speak to one another to build at least tentative alliances against developmentalism.

Highlighting that contemporary land grabbing is deeply entangled with global environmental crises, in Chapter 3, *Felix Mantz* analyzes the employment of coercive technologies to undermine struggles against land grabbing in Tanzania. This analysis is to a significant extent based on research conversations with Tanzanian Indigenous organizers, and pastoral activists. In particular, Mantz examines three institutions that are employed to suppress anti-land grabbing resistance: prisons, borders, and police, which are mobilized to advance land grabbing for conservation purposes. Indeed, much of the organizing against land grabbing in Tanzania must contend with these coercive institutions that expose communities to direct forms of violence in the state and capital's attempt to gain power over lands. Thus, those who seek to be in active solidarity with struggles against land grabbing must ask (1) how can we enhance frontline communities' ability to defend against and deter the use of direct violence, and (2) how do we weaken the capabilities of state and capital to unleash coercive technologies of direct violence?

Chapter 4 features a conversation between *Samwel Nangiria*, a prominent Maasai activist from northern Tanzania and co-founder of the Pan African Living Culture Alliance, and *Nick Lunch*, the co-founder and director of a participatory development organization (InsightShare) in solidarity with Indigenous peoples' struggles for self-determination. Reflecting on the Maasai struggle against colonial conservation, land grabbing, and cultural erasure in northern Tanzania, this conversation discusses the revolutionary possibilities of participatory video (PV). Specifically, PV has been used by activists in Tanzania to report and highlight violence against Indigenous peoples in the name of conservation that would have otherwise gone unwitnessed. It has also been employed to establish radical alliances, collectives, and networks of knowledge exchange across Africa and beyond. Finally, Nangiria and Lunch discuss how PV can strengthen Indigenous and oppressed communities' sense of unity, shared struggle, and politicization. Most centrally, they argue that PV enables

Indigenous peoples to tell their own stories, an exercise in self-determination that has influenced transnational processes—from COP26 to efforts of decolonizing British museums. The authors propose PV as a tool to facilitate the systemic shift from "colonial conservation to ecological decolonization."

In Chapter 5, *Ali Kaba* and *Baba Sillah* examine local communities' responses against top-down forest conservation regimes in Liberia. While the authors provide an overview of the state of land grabbing in Liberia, they focus particularly on the Kranh-Bassa Proposed Protected Area (KBPPA) in the southeastern part of the country. They identify growing discontent among local communities about the threat KBPPA poses to their sense of identity and livelihoods, as well as the lack of viable alternative livelihoods amid land dispossession. The chapter tells a story of developing and adapting strategies of resistance whereby communities initially sought to petition the state to take their interests seriously. However, following growing disillusionment with agreed upon rules, the absence of legal clarity, and the lack of any real change to the conditions on the ground, more confrontational strategies were employed. This included a refusal to cooperate and recognize specific rules as well as forms of sabotage and even direct confrontation with park rangers. A key insight that emerges from Kaba and Sillah's analysis is the reliance (and vulnerability) of state institutions on the cooperation of local communities to realize control over lands and forests.

In Chapter 6, *Mariko Frame* presents interviews with Cambodian women who were evicted when the land surrounding Boeung Kak Lake, and indeed even the lake itself, was appropriated for development. Highlighting the gendered aspects of land grabbing, the interviews illustrate how women are active participants and critical leaders in resistance against land grabbing. However, as other reports and studies concerning land grabbing in Cambodia show, while women are important leaders, they suffer from arrests and violence at the hands of security forces. Lacking substantive equality with men, who also hold the majority of land titles, women face severe socioeconomic problems in the face of land grabbing, including loss of livelihood and assets, political repression, psychological and physical harm, and deteriorating familial relations. As elsewhere in the Global South, women are active and crucial actors in movements of resistance, yet they continue to pay a heavy toll from the intersectionality between patriarchy, class, and the dynamics of capitalist accumulation.

Chapter 7 is a dialogue between the editors and Cambodian activist and former project director of Prey Lang Community Network (PLCN), *Narith Nou*. Prey Lang is a major rainforest in northern Cambodia, a biodiversity hotspot, and home to a quarter of a million inhabitants. In the interview, Nou details the struggles by organizers against the ongoing deforestation of Prey Lang through industrial agriculture, illegal logging, and mineral extractivism. He tells a story of violent state repression in collusion with domestic and international corporate interests, and the efforts by PLCN to organize and mobilize affected communities. The interview provides insights into decentralized organizational movement structures such as those employed by PLCN as well as the difficult and often life-threatening conditions of repression and violence under which its members work and agitate for change. Nou illustrates what "development" through forest grabbing really means for Prey Lang: displacement, dispossession, and poverty. Crucially, the interview elucidates the deep and inseparable entanglement of struggling to protect poor people and struggling to protect ecosystems such as rainforests. Fighting deforestation is not only a fight for biodiversity or conservation but also for the improvement of those who live in and from forests.

Pluriversal Perspectives for and beyond Resistance

In Chapter 8, *Leonardo Figueroa Helland, Angela Martinez, Neftalí Reyes Mendez, Angélica Castro Rodríguez, Juan José López Negrete, Mileida Correa, José Gualinga*, and *Çaca Yvaire* give voice to Indigenous communality and reciprocity in biocultural territories of life. The authors discuss four key processes from Abya Yala and Turtle Island that embody decolonizing communal horizons of regenerative emancipation: (1) The Kichwa People of Sarayaku's Kawsak Sacha/Living Forest (Ecuador); (2) The Association of Indigenous and Afrodescendant Fishermen and Peasants for the Community Development of the Bajo Sinú (Colombia); (3) EDUCA, A.C., Services for an Alternative Education's campaign for Communal Alternatives in the Defense of Territory (Mexico); and (4) NEFOC, Northeast Farmers of Color Land Trust (Turtle Island/United States). These processes embody decolonizing solutions to intersecting crises. Rooted in land-based communal self-determination, they revitalize Indigenous knowledge and governance. Upon Indigenous cosmovisions, they (re)weave sacred territories where biocultural diversity, food sovereignty, and political agroecology enact autonomous multispecies communities. Furthermore, these processes are committed to strengthening an intersectional community fabric that celebrates gender diversity and nurtures Indigenous-Black-peasant, people of color, and broader intersectional solidarities in struggles for systemic change. The chapter argues that Indigenous resurgence processes move us toward territorially sovereign, anticolonial, anti-patriarchal, and noncapitalist forms of communal governance that actualize "pluriversal" worlds.

Chapter 9, written by *Samuel L. Grant*, draws on Africana critical theory to examine the liberation praxis of various movements primarily across the African continent. These movements encompass three broad groups: African Indigenous peoples, African peasants, and African slum dwellers. In particular, the chapter scopes a wide variety of organizations, formations, processes, and struggles that are "all planting seeds of liberation." It demonstrates how these formations resist land dispossession, poverty and deprivation, cultural erasure, and food insecurity produced and reproduced by ecological imperialism. The chapter argues that the next stage of struggle requires the co-organizing of these various formations, across Indigenous, peasant, and slum survivors' movements. This is because, at the moment, there is not sufficient "cro-creative liberation praxis" that could pose a sustained threat to ecological imperialism.

In Chapter 10, *Rose Brewer* analyzes the current state of global ecological destruction and social inequalities from the perspective of the "African world," including the African continent, the African diaspora, the Black United States, and Afro-descendants across the globe. From this perspective, she names the current world order "imperial racial monopoly capitalism" and examines its multiple dimensions and dynamics. Particular attention is paid to the gendered logics of this world order and women's struggles. The chapter surveys several instances of resistance and uprising against imperial racial monopoly capitalism—from the Garifuna struggle in Honduras to Cooperation Jackson in the United States. Ultimately, the chapter calls for "organizing as imperative." This mode of organizing is based on at least two key pillars: international solidarity to link and co-organize across struggles, and deep relationality among humans and nonhumans.

Chapter 11 by *Ali Kassem* identifies and explores the existence of a lived Muslim cosmology and mode of being that cannot be squared with anthropocentrism, modernity, and coloniality. Kassem' analysis is based on his long-standing research in Lebanon. Among others, this research features a series of conversations with Muslim students and activists who live in Beirut where they experience an urban alienation and estrangement from nature. Reflecting on this experience alongside and with these conversations, Kassem identifies strands of Shia Muslim lived cosmologies and epistemologies that, while existing at the margins of wider Shia, Muslim, or Lebanese societies, hold the potential for non-anthropocentric relations with ecologies, animals, and the spiritual. Kassem introduces a wide range of entry points to further explore, develop, and manifest this potential. These include, among others, the principle of cosmic unity and balance, spiritual journeys toward synchronicity with the universe, non-anthropocentric understandings of rights-bearers, lessons drawn from Shia mystics and teachings by the Prophet, and questions of ethical relationships to food.

In Chapter 12, *Abigail Pérez Aguilera* reflects on the possibilities of moving toward multispecies resistance. Drawing on affectivity and radical decolonial imaginaries, the chapter emphasizes the need to move beyond hegemonic ways of relating to the nonhuman and more-than-human world. Such a relationality, Aguilera argues, requires the recognition that our conventional understanding of the human and "nature"—as well as our ideas of how to struggle and resist in the face of climate and environmental crises—is deeply conditioned by Euro- and anthropocentric epistemologies. To undo this, the chapter emphasizes the need to find "fractured loci," spaces within coloniality where resistance and different relationalities that include the nonhuman can take place. The chapter considers examples from Mexico—environmental conflict in Wirutuka and Mexico City's resistant presence of water bodies and rivers in the face of modernization and development—to illustrate how forms of multispecies resistance, and struggles from a non-anthropocentric and decolonial fractured locus can look like.

In the final chapter, the editors reflect on key lessons that can be learned from the cases in this book. They highlight three insights in particular, which arise from considering the chapters all together. The first insight is the highly stratified consequences of extractivism and the pluriverse of responses they elicit. Extractivism manifests differently depending on specific contexts, their histories, and ecological and social, political, and economic relations. Likewise, responses to extractivism are highly diverse. The second insight which emerges concerns the various, complex relations to the state. The state often appears as the central agent that movements immediately have to confront in their efforts of resistance against extractivism. Many of these movements have visions of alternative socio-ecological futures autonomous from the state. However, regarding strategy and tactics, the state has to be engaged with in various ways and capacities. Finally, the cases also illustrate the need for pluriversal, international solidarity in the face of global capitalism. While place-based successes can bring significant changes for the lived realities of the peoples involved, all communities face the continued threat of global capitalism and extractivism as long as the modern, colonial, capitalist, and patriarchal international order remains in place.

Bibliography

Alimonda, Héctor. 2019. "The Coloniality of Nature: An Approach to Latin American Political Ecology." *Alternautas* 6(1): 102–42.

Amin, Samir. 1994. "The Future of Global Polarization." *Review (Fernand Braudel Center)* 17(3): 337–47.

ASEAN. n.d. "ASEAN Free Trade Area Agreements." *ASEAN Investment*. Retrieved July 11, 2023. https://investasean.asean.org/asean-free-trade-area-agreements/view/757/newsid/872/asean-trade-in-goods-agreement.html

Bhambra, Gurminder K. 2018. "The State—Postcolonial Histories of the Concept." In Olivia U. Rutazibwa and Robbie Shilliam (eds.), *Routledge Handbook of Postcolonial Politics*, 200–9. Abingdon: Routledge.

Campbell, Bonnie and John Loxley. 1989. *Structural Adjustment in Africa.* New York: St. Martin's Press.

Columbus, Christopher. 2003 "Journal of the First Voyage of Columbus." In Julius Olson and Edward G. Bourne (eds.), *The Northmen, Columbus and Cabot, 985–1503*, 85–258. New York: Charles Scribner's Sons.

Dorninger, Christian, Alf Hornborg, David J. Abson, Henrik von Wehrden, Anke Schaffartzik, Stefan Giljum, John-Oliver Engler, Robert L. Feller, Klaus Hubacek, and Hanspeter Wieland. 2021. "Global Patterns of Ecologically Unequal Exchange: Implications for Sustainability in the 21st Century." *Ecological Economics* 179: 106824.

Dunlap, Alexander. 2021. "The Politics of Ecocide, Genocide and Megaprojects: Interrogating Natural Resource Extraction, Identity and the Normalization of Erasure." *Journal of Genocide Research* 23(2): 212–35.

Economics in Context Initiative. 2021. "Comparative Economic Systems: Capitalism and Socialism in the 21st Century." *An ECI Teaching Module on Social and Economic Issues.* Global Development Policy Center, Boston University.

Fanon, Frantz. 2011. *The Wretched of the Earth.* New York: Grove Press.

Federici, Silvia. 2004. *Caliban and the Witch: Women, the Body and Primitive Accumulation.* Brooklyn: Autonomedia.

Figueroa Helland, Leonardo and Tim Lindgren. 2016. "What Goes around Comes Around: From the Coloniality of Power to the Crisis of Civilization." *Journal of World-Systems Research* 22(2): 430–62.

Figueroa Helland, Leonardo, Abigail Pérez Aguilera, and Felix Mantz. 2021. "Decolonize, ReIndigenize: Planetary Crisis, Biocultural Diversity, Indigenous Resurgence, and Land Rematriation." In Suzanne McCullagh, Luis Prádanos, Ilaria Tabusso Marcyan, and Catherine Wagner (eds.), *Contesting Extinctions: Decolonial and Regenerative Futures*, 23–62. Lanham: Lexington Books.

Forests and Finance. n.d. "Forests & Finance: Home." Retrieved July 12, 2023. https://forestsandfinance.org/

Frey, Scott, Paul K. Gellert, and Harry F. Dahm. 2018. *Ecologically Unequal Exchange: Environmental Injustice in Comparative and Historical Perspective.* Cham: Palgrave MacMillan.

Frame, Mariko L. 2023. *Ecological Imperialism, Development, and the Capitalist-World System: Cases from Africa and Asia.* London: Routledge.

Francis, Romain. 2019. "The Tyranny of the Coloniality of Nature and the Elusive Question of Justice." In Everisto Benyera (ed.), *Reimagining Justice, Human Rights and Leadership in Africa: Challenging Discourse and Searching for Alternative Paths*, 39–57. Cham: Springer.

Giljum, Stefan and Nina Eisenmenger. 2004. "North-South Trade and the Distribution of Environmental Goods and Burdens: A Biophysical Perspective." *Journal of Environment and Development* 13(1): 73–100.

Grimes, Peter and Jeffrey Kentor. 2003. "Exporting the Greenhouse: Foreign Capital Penetration and CO? Emissions 1980–1996." *Journal of World-Systems Research* 9(2): 261–75.

Grosfoguel, Ramón. 2013. "The Structure of Knowledge in Westernized Universities-Epistemic Racism/Sexism and the Four Genocides/Epistemicides of the Long 16th Century." *Human Architecture: Journal of the Sociology of Self-Knowledge* 11(1): 73–90.

Harvey, David. 2007. *A Brief History of Neoliberalism.* New York: Oxford University Press.

IGBP. n.d. *Great Acceleration.* IGBP. Retrieved July 12, 2023 http://www.igbp.net/ globalchange/greatacceleration.4.1b8ae20512db692f2a680001630.html

Jorgenson, Andrew K. 2007. "Does Foreign Investment Harm the Air We Breathe and the Water We Drink? A Cross-national Study of Carbon Dioxide Emissions and Organic Water Pollution in Less Developed Countries, 1975–2000." *Organization & Environment* 20(2): 137–56.

Jorgenson, Andrew K. 2009. "The Transnational Organization of Production, the Scale of Degradation, and Ecoefficiency: A Study of Carbon Dioxide Emissions in Less Developed Countries." *Human Ecology Review* 16(1): 64–74.

Jorgenson, Andrew K. 2010. "World-Economic iIntegration, Supply Depots, and Environmental Degradation: A Study of Ecologically Unequal Exchange, Foreign Investment Dependence, and Deforestation in Less Developed Countries." *Critical Sociology* 36(3): 453–77.

Klare, Michael. 2012. *The Race for What's Left: The Global Scramble for the World's Last Resources.* New York: Picador.

Krausmann, Fridolin, Fischer-Kowalski, Marina, Schandl, Heinz., and Eisenmenger, Nina. 2008. "The Global Sociometabolic Transition Past and Present Metabolic Profiles and their Future Trajectories." *Journal of Industrial Ecology* 12(5–6): 636–56.

Lugones, María. 2007. "Heterosexualism and the Colonial/Modern Gender System," *Hypatia* 22(1): 186–219.

McMichael, Philip. 2017. *Development and Social Change: A Global Perspective*, 6th ed. Thousand Oaks: Sage Publications.

Mignolo, Walter and Catherine Walsh. 2018. *On Decoloniality: Concepts, Analytics, Praxis*. Durham: Duke University Press.

Moore, Jason. 2017. "The Capitalocene, Part I: On the Nature and Origins of Our Ecological Crisis." *The Journal of Peasant Studies* 44(3): 594–630.

Ndlovu-Gatsheni, Sabelo J. 2015a. *Empire, Global Coloniality and African Subjectivity*. New York: Berghahn Books.

Ndlovu-Gatsheni, Sabelo J. 2015b. "Genealogies of Coloniality and Implications for Africa's Development." *Africa Development* 40(3): 13–40.

Ndlovu-Gatsheni, Sabelo J. 2018. "Against Bringing Africa 'Back-In.'" In Marta Iñiguez de Heredia and Zubairu Wai (eds.), *Recentering Africa in International Relations: Beyond Lack, Peripherality, and Failure*, 283–306. Palgrave Macmillan.

Nişancıoğlu, Kerem. 2020. "Racial Sovereignty." *European Journal of International Relations* 26(S1): 39–63.

Organization for Economic Development and Cooperation (OECD). 2002. "Foreign Direct Investment for Development: Maximising Benefits, Minimising Costs." *OECD*. https://www.oecd.org/investment/investmentfordevelopment/foreigndirectin vestmentfordevelopmentmaximisingbenefitsminimisingcosts.htm

Parrique, T., Barth, J., Briens, F., Kerschner, C., Kraus-Polk, A., Kuokkanen, A., and Spangenberg, J. H. 2019. *Decoupling Debunked: Evidence and Arguments Against Green Growth as a Sole Strategy for Sustainability*. European Environmental Bureau (EEB) & Make Europe Sustainable for All. https://mk0eeborgicuypctuf7e.kinstacdn .com/wp-content/uploads/2019/07/Decoupling-Debunked.pdf

Perez-Rocha, Manuel. 2021. "Missing from the Climate Talks: Corporate Powers to Sue Governments Over Extractives Policies." *Inequality.org*. https://inequality.org /research/missing-from-the-climate-talks-corporate-powers-to-sue-governments -over-extractives-policies/

Plumwood, Val. 1993. *Feminism and the Mastery of Nature*. London: Routledge.

Querejazu, Amaya. 2016. "Encountering the Pluriverse: Looking for Alternatives in Other Worlds." *Revista Brasileira de Política Internacional* 59(2): e007.

Quijano, Aníbal. 2000. "Coloniality of Power, Eurocentrism, and Latin America." *Nepantla: Views from South* 1(3): 533–80.

Quijano, Aníbal. 2007. "Coloniality and Modernity/Rationality." *Cultural Studies* 21(2–3): 168–78.

Rodney, Walter. 2018. *How Europe Underdeveloped Africa*. London: Verso.

Shandra, John M., Michael Restivo, and Jamie M. Sommer. 2019. "Appetite for Destruction? China, Ecologically Unequal Exchange, and Forest Loss." *Rural Sociology* 85(2): 346–75.

Steffen W., A. Persson, L. Deutsch, J. Zalasiewicz, M. Williams, K. Richardson, C. Crumley, P. Crutzen, C. Folke, L. Gordon, M. Molina, V. Ramanathan, J. Rockström, M. Scheffer, H. J. Schellnhuber, and U. Svedin. 2010. "The

Anthropocene: From Global Change to Planetary Stewardship." *Ambio* 40(7): 739–61.

Thiong'o, Ngũgĩ wa. 2011. *Decolonising the Mind: The Politics of Language in African Literature.* Oxford: James Currey.

Transnational Institute (TNI). 2013. *The Global Land Grab: A Primer.* Amsterdam: TNI. https://www.tni.org/files/download/landgrabbingprimer-feb2013.pdf

United Nations Conference on Trade and Development (UNCTAD). 2005. *Economic Development in Africa: Rethinking the Role of Foreign Direct Investment.* New York: United Nations.

United Nations Conference on Trade and Development (UNCTAD). 2012. *Economic Development in Africa Report 2012: Structural Transformation and Sustainable Development in Africa.* New York: United Nations. https://unctad.org/system/files/official-document/aldcafrica2012_embargo_en.pdf

Vázquez, Rolando. 2017. "Precedence, Earth and the Anthropocene: Decolonizing Design." *Design Philosophy Papers* 15(1): 77–91.

Wolfe, Patrick. 2006. "Settler Colonialism and the Elimination of the Native." *Journal of Genocide Research* 8(4): 387–409.

Wynter, Sylvia. 2003. "Unsettling the Coloniality of Being/Power/Truth/Freedom: Towards the Human, After Man, Its Overrepresentation-An Argument." *CR: The New Centennial Review* 3(3): 257–337.

Yu, Yang, Kuishuang Feng, and Klaus Hubacek. 2014. "China's Unequal Ecological Exchange." *Ecological Indicators* 47: 156–63.

Coming Home

Haider Khan

COMING HOME (Manila, the Philippines) (A poem dedicated to the memory of the revolutionary artist, journalist, and freedom fighter, the late Kamal Lohani) Kahnapad Haider (Kahnapad is the name of a ninth- to tenth-century Tantric Buddhist poet from Bengal).

………..N O………..

this is not my home wasn't

meant to be

these neon signs dammed-up traffic so many monsters

in a Developmental Package

these

glamorous

third world prostitutes

their

foreign patrons

this antiseptic
closed workplace

of

masks
and
masquerades

outside in the bus stands people wait

in the rain

No...
The romance of rain of my distant childhood
No more Only the mud-soaked

Sewers Global Capital

reaches

Everywhere

Tokyo Paris Persepolis Timbaktu New York New Delhi

The Capitals und Das Kapital

dance a frenzied multinational free-for-all unreal dance
all these TNCs masters of their states and fates
of dark people far away

No.....
these tourist brochures of

Boracay Baguio Cebu or Bohol

Merely titillate
Senses that are no longer natural

Happiness on an installment plan easy to buy on cash or credit

life

Is no more complex
than a work-out plan devised by
Workout Artists

With the proper credentials
from Ivy League schools

A mixed-breed lingo on TV recalls Caliban muttering about Prospero's power
Crawling at the bottom of the urban sea

Bonifacio's ghost hovers somewhere along with Soekarno and Nehru and
N'Kruma and Lumumba

Mao 's continuous revolution halted one hopes only for a time
Kalayan Katipunan etc. etc.
on the national flag recalls memories of a revolt buried
in the deep dark corners
In the burial ground where all dangerous memories
Lie entombed

Global finance hovers like a goblin
in a helicopter that charges several thousand dollars an hour the white
dignitaries from afar look down on the traffic jam below.

No......

no more the rain falling on hyacinths all night long
or the lightning in the dark eyes of my sister
or the scent of the flower in her hair only dead remembrances alive in anxious
guilt

No more

dreams of a vibrant childhood

But who will remember-----
The floods the traffic the prostitutes?
And those screaming staring hungry children's eyes?

(H. A. Khan [aka Kahnapad Haider],
Distinguished University Professor, JKSIS, University of Denver)

II

Existing Struggles

1

The Pathalgadi Movement of Jharkhand, India

Three Theses on Indigenous Rebellion at the Limits of Liberal Democratic Politics

Pratik Raghu

From 2017 to 2019, predominantly adivasi (South Asian Indigenous) communities in Jharkhand, India, put up large stone slabs known as pathals at their village entrances. These slabs reaffirmed their independence from the Jharkhandi and Indian states and their opposition to the extraction of precious resources on their lands by corporations and state authorities. Participating communities established their own self-defense committees, banks, and schools, defying calls for assimilation through the neoliberal Hindu nationalism that has proliferated in the state since it achieved independent statehood in 2000. The Pathalgadi Movement, as it became known, challenged not only state sovereignty and corporate extractivism but also Indigenous politics trapped within the domains of the state, capital, and civil society.

In spite of its audacious experiments with autonomy, the Pathalgadi Movement was framed as little more than an exercise in constitutionalism by liberal commentators. In this chapter, I show how the Pathalgadi Movement exceeded the terms of liberal democracy by putting forward three theses that should provoke scholar-activists who seek solidarity with Indigenous and other oppressed grassroots communities to reexamine their foundational assumptions and standard practices. More specifically, I contend that 1) ecological imperialism often intertwines with domestic capitalism and national chauvinism; 2) Indigenous communities are frequently riven by class differences that can produce political dissensus; and 3) these communities not uncommonly challenge liberal democratic institutions at the same time as they strategically engage them, calling for a novel conceptual apparatus that I call a repertoire of "subaltern ungovernability." Failing to reckon with non-liberal or even anti-liberal emancipatory politics in this vein and contrarily situating the communities in question squarely within the liberal logics of governmentality and sovereignty is analytically unsound, socially irresponsible, and politically dangerous.

From its inception in early 2017 to its suppression in mid-2019, the Pathalgadi Movement for communal autonomy, led by the sizable Munda Indigenous community in the eastern Indian state of Jharkhand, defied easy categorization by sympathizers and opponents alike. Instigated by the right-wing Hindu nationalist Bharatiya Janata Party (BJP)'s attempts to override long-standing land laws that prevent the transfer of tribal land to non-tribals, the movement erected large stone slabs known as pathals at the entrances to primarily adivasi communities in Jharkhand and its neighboring states of Chhattisgarh and Odisha.[1] Movement participants listed constitutional and legal protections guaranteed to adivasis on these stone slabs, subsequently using them to declare their independence from the Indian and Jharkhandi states and develop their own community-based schools, banks, and defense patrols. Jharkhand's BJP-controlled state government at the time vilified the movement in every possible way, variously blaming Maoist insurgents, Christian missionaries, drug traffickers, and other outside agitators for manipulating "poor, innocent tribals." At the same time, prominent liberal activists, journalists, intellectuals, and other civil society actors in Jharkhand and wider India struggled to situate the Pathalgadi Movement, in all its complexity and contradiction, within a purely pacifist, constitutionalist, or rights-based framework.[2] A number of these actors admirably set aside their misgivings to offer support to movement participants who were arrested en

---

[1] The term "adivasi" means "inhabitant since the beginning" in Hindi, and it is the collective name commonly used by many Indian and more broadly South Asian Indigenous communities. Though Indigenous peoples in India are officially recognized as "Scheduled Tribes," this is a legal and constitutional term imposed upon the populations at hand. The term "adivasi" contrarily emerged from political mobilizations to form a sense of pan-Indigenous identity in the 1930s, in no small part by affirming Indigenous claims to land and land-based self-governance systems (Minority Rights Group International, 2022). Mundas make up one of India's largest adivasi populations: they speak the Mundari language and primarily reside in the states of Jharkhand, Odisha, and West Bengal, as well as in portions of Nepal and Bangladesh.

[2] For the purposes of this chapter, I define liberalism as a Eurocentric political philosophy centered around civil liberties, individual rights, representative government, and free enterprise. It emerged in the eighteenth and nineteenth centuries as the ideology of emerging capitalism, and it continues to serve that function even as the neoliberal variant of capitalism becomes increasingly plagued by crises.

masse by the Jharkhandi state government in mid-2019, but their unease about the movement's ideology and methods has by no means necessarily abated.

The analytical and political uncertainty that the Pathalgadi Movement generated actually offers valuable insights into the complexities of Indigenous communities and the movements they articulate in India and beyond. Scholars, activists, and policymakers who engage these communities and movements all too often operate within a false dichotomy that depicts Indigenous peoples as innate, unequivocal stewards of nature, on the one hand, and walking anachronisms who will be inevitably absorbed by modernization, on the other. In actuality, many Indigenous communities operate at the limits of late capitalist modernity: they strategically engage the state, capital, and civil society while simultaneously striving for autonomy. Due in no small part to their entanglements with these hegemonic domains, Indigenous communities have frequently had to adapt their ecological traditions to suit the social, political, and economic circumstances imposed upon them, which is to say that Indigenous mobilizations do not simply extend preexisting customs into the present. None of these nuances extinguish the emancipatory potential of Indigenous politics as it is practiced in numerous locales. However, recognizing this potential calls for a reckoning with the exigencies of neoliberalism, neo-imperialism, neocolonialism, and neofascism that numerous Indigenous communities confront as they determine how to engage the state, capital, and civil society.

In this chapter, I use the case of the Pathalgadi Movement to put forward three theses about Indigenous rebellion at the limits of liberal democratic politics. These theses should be tested, so to speak, against other instances of Indigenous rebellion against extractivism, dispossession, and displacement that exceed the terms of liberal democracy. In advancing these theses, I show how the Pathalgadi Movement has strategically engaged, disengaged, and deceived the state, articulating what I call a repertoire of subaltern ungovernability.

My theses are grounded in my dissertation research on the Pathalgadi Movement, for which I extensively surveyed English-language news reports and editorials about the Movement, studied a number of monographs and scholarly articles detailing the histories of Indigenous rebellion in Jharkhand, and conducted a total of eighteen semi-structured interviews with Jharkhand-based nonprofit professionals, academics, activists, and journalists between 2018 and 2019. Even though the Pathalgadi Movement made national headlines across India at its peak, scholarly interventions analyzing its rationale, dynamics, and outcomes—at least in the English language—remain relatively scarce. I can only ultimately speculate about the reasons for this paucity of academic attention, but most existing scholarship on the movement reaffirms the legalistic postcolonial liberal paradigm I critique here, potentially illustrating the latter's self-fulfilling feedback loops. In other words, liberalism may well have sanded off the ideologically inconvenient aspects of the Pathalgadi Movement—those edges that poke through established legal and constitutional parameters—through the sheer force of repetition.

On a more practical level, the lack of attention accorded to the Pathalgadi Movement could also highlight the partial success of the increasingly draconian Indian state, under the direction of the Hindu Right, in preventing widespread awareness of organized challenges to its neoliberal ethno-nationalist agenda. On the other hand, the Pathalgadi Movement's limited resonance with the emergent international politics of Indigeneity may concomitantly cast the latter's typically urban, middle-class, educated trappings in sharper relief. My second thesis draws upon the work of Alpa Shah to explore this unsettling but crucial limitation in slightly greater depth. My greater reliance on publicly available reports and commentaries as well as my own interviews and observations thus reflects the aforementioned academic literary gap at the same time as this chapter attempts to address it.

Drawing upon my own past misconceptions and mistakes, my interventions more so address non-Indigenous scholar-activists who seek to work with Indigenous communities than the communities in question themselves. The Indigenous and other grassroots resistances that this volume highlights will struggle to foster broad-based liberation if supposedly sympathetic scholars continue to understand these mobilizations solely through the restrictive and perhaps even self-defeating lens of liberal democratic politics.

Thesis 1. Ecological imperialism often intertwines with domestic capitalism and national chauvinism, forcing Indigenous and other oppressed grassroots communities to fight multidimensional battles to protect their lands, livelihoods, and cultures.

Extractivism in Jharkhand is at once a regional, national, and transnational endeavor. The state is richly endowed with coveted natural resources such as iron ore, coal, mica, bauxite, uranium, and limestone, accounting for nearly 40 percent of India's total mineral reserves (S. Singh, 2018). Particularly since the liberalization of the Indian economy in 1991, various private corporations from India as well as many other parts of the world, in addition to public entities such as the Uranium Corporation of India, have striven to capitalize on these abundant endowments (Xaxa, 2018). Virtually regardless of their political affiliations, state authorities have willingly facilitated the corporate expropriation of Jharkhand's resources since the region achieved independent statehood in 2000. In May 2016, the BJP government of Chief Minister Raghubar Das amended the 1908 Chotanagpur Tenancy Act (CNTA) and the 1949 Santhal Parganas Tenancy Act (SPTA) to enable the seizure of 2.1 million acres of land that had housed sacred groves, village paths, graveyards, and other local adivasi community institutions (Dungdung, 2022a). Das and the BJP made their plans for this land clear in February 2017 when they hosted the Momentum Jharkhand Global Investors Summit in Ranchi. During this summit, the Jharkhand State Government signed 210 memoranda of understanding worth 3.12 trillion rupees (over 37 billion dollars) with various corporations, with the majority of proposals being in the mining and geology sectors (Majumdar, 2017). These combined developments provided a powerful impetus for the rise of the Pathalgadi Movement.

The Adani Group—a multinational conglomerate of Indian origin that secured the right to produce urea, methane, power, and substitute natural gas during the Momentum Jharkhand Summit—exemplifies not only how Indian capitalists have sought to make inroads into Jharkhand but also how these capitalists in and of themselves belong to the transnational capitalist class. As such, many of India's economic elites are not simply compradors but protagonists in extractivism, dispossession, and displacement both at home and abroad. While India might arguably fall short of imperialist or even sub-imperialist status at present, its ruling class's economic activities are not necessarily domestic even when they take place on home soil. The raw materials extracted by Indian multinational corporations from a locale like Jharkhand fit into patterns and mechanisms of private capital accumulation that far exceed India's borders. For example, the Adani Group—whose CEO Gautam Adani is one of incumbent BJP Prime Minister Narendra Modi's biggest corporate backers—is attempting to construct Australia's largest ever coal mine in Queensland. These dynamics complicate simplistic depictions of imperialism as a burden purely or at least primarily foisted by the Global North upon the Global South, though these complexities by no means absolve actors from the Global North of their culpability.

In addition to blurring the lines between ecological imperialism and domestic capitalism, the dynamics of extractivism in Jharkhand also demonstrate how neoliberal globalization can articulate with ultra-parochial nationalism. The BJP is the public-facing electoral organ of the Sangh Parivar or family of Hindu nationalist organizations, which broadly seeks to supplant constitutional secularism with the upper-caste Hindu control of India's governing institutions, resources, citizenry, and borders. The rise of Hindu nationalism coincided and ultimately dovetailed with the liberalization of India's economy in the 1990s, even though this process was initiated and has been just as much embraced by the centrist Indian National Congress (INC), the BJP's primary political rival. The Hindu Right has a long-standing presence in Jharkhand: the BJP prioritized Jharkhandi statehood to court support among adivasis, proposing Hinduization alongside neoliberalization as the ultimate solutions to the supposedly innate "backwardness" of adivasi populations (Shahdeo, 2022). Meanwhile, the Rashtriya Swayamsevak Sangh (People's Voluntary Organization or RSS for short)—a paramilitary grassroots organization that serves as the central coordinating body of the Sangh Parivar—has attempted to slowly and surreptitiously indoctrinate and incorporate Jharkhandi adivasi communities while providing education, health, and other much-needed social services for decades. Persuading adivasis to abandon their traditional land-based beliefs and practices is pivotal to persuading them to abandon their traditional lands, leaving the latter open to private and public enclosure.

The Das regime exemplified the interconnectedness of ecological imperialism, domestic capitalism, and Hindu nationalism. In confronting the hydra of neoliberal Hindu nationalism, Pathalgadi Movement participants rejected "false *sanskriti*" (upper-caste Hindu culture) at the same time as they prevented state and capitalist power brokers from physically occupying and depleting their mineral-rich lands (Sundar, 2018). While the influence of Hindu nationalism might be particular to Jharkhand and India, academics and activists who seek to help Indigenous and other oppressed grassroots communities fight ecological imperialism would do well to recognize how neoliberalism operates differently in different contexts. They must avoid treating neoliberalism either as a standardized modality of hegemony and domination, or as the lone factor in extractivism, dispossession, and displacement across the world. They must contrarily reckon with neoliberalism's context-specific political, economic, and cultural entanglements if they hope to support their interlocutors in broad-based struggles for liberation as opposed to single-issue campaigns.

Thesis 2. Indigenous communities and sites of struggle are frequently riven by class differences that can produce political dissensus and thus complicate the cultivation of solidarity.

Researchers and organizers striving for solidarity with Indigenous communities must grapple not only with the complexities of external oppressive forces but the internal contradictions of the communities at hand. Like all other oppressed populations, Indigenous communities are not monolithic, with class serving as an especially potent source of politically consequential internal divisions. Alpa Shah chronicles how predominantly educated, urban, middle-class Indigenous activists in Jharkhand have unwittingly subjected predominantly uneducated, rural, and poor and working-class adivasi communities to "eco-incarceration." Determined to preserve the romanticized image of adivasis as immutable ecological guardians living in harmony with nature, these relatively privileged activists have leveraged state power to prevent these communities from killing animals that endanger their safety and from temporarily migrating to urban areas for extra work to support their families (Shah, 2010: 113–16, 138–43). As such, they have effectively trapped or "incarcerated" these communities in their respective environments. Needless to say, eco-incarceration is an untenable solution to ecological imperialism. Aside from harming the very communities it purports to protect, eco-incarceration's frequently depoliticized conceptualization of culture does not, in itself, act as a bulwark against extractivism, dispossession, and displacement.

Depoliticizing cultural discourses can breed hostility toward Indigenous communities who interpret and act upon their traditions in far more militant ways, as exemplified by Jharkhandi civil society's general responses to the Pathalgadi Movement. When I first visited Jharkhand in the summer of 2018, the state was still reeling from the disruption caused by the movement, as evinced by the deployment of security forces to Ranchi, the state capital, and Khunti, a rural town about thirty kilometers from Ranchi that was one of the movement's epicenters. Despite the looming danger of state violence, virtually all of the nonprofit professionals that I interviewed in Khunti—the majority of whom are Munda adivasis, just like the majority of Pathalgadi participants— denounced the movement's subversion of adivasi custom, its non-compliance with state and national law, and, in turn, its refusal to endorse the neoliberal civil society agenda of cultural preservation, development, and good governance. Some of these professionals acknowledged that the movement's grievances about the governmental and corporate seizure of land were at least partly legitimate and that adivasis generally desire and perhaps even deserve self-determination. However, they followed these tentative concessions with rather acerbic criticisms of movement participants for politicizing and thereby sullying a sacred adivasi tradition, excluding well-meaning outsiders, allegedly indulging in alcoholism and human trafficking, and, more generally, for not recognizing the futility of their anti-statist actions. Certain liberal academic commentators echo this note of futility, lamenting the Pathalgadi Movement's propagation of "half-baked schemes . . . oblivious to the nuances of the present politics" (A. Singh, 2019: 31–2).

The professionals at hand, who worked at the time for an array of local, national, and international nongovernmental organizations, juxtaposed the movement's opposition to well-meaning outsiders, such as themselves as well as state bureaucrats and welfare service providers, with its alleged links to malicious outsiders, especially "left-wing extremists." Even more striking than these unprovable but nonetheless pointed accusations was the claim made by multiple professionals that adivasis have the right to protect their own cultures at the same time as they do not have the right to construct "parallel institutions," such as the schools, banks, and other autonomous initiatives set up by the Pathalgadi agitators. This self-contradictory statement reveals that, from the perspective of the nongovernmental sector that has come to dominate civic associational life in Jharkhand, the Pathalgadi Movement does not only lie outside the boundaries of civil society but is further opposed to the latter's rational, respectable, and modern liberal sensibilities, for which it must be taken to task.

The Jharkhandi activists and intellectuals with whom I discussed the movement—again, a mix of adivasis and non-adivasis—were more overtly sympathetic to its intentions while nonetheless reiterating some of the aforementioned criticisms. While they too did not approve of the movement's divergence from the older, less confrontational Munda practice of Pathalgadi— which has been used to commemorate ancestors, clarify village rules, demarcate village boundaries, display a community's history, and memorialize special events (Sahu, 2018; Dundgung, 2022a)—their objections focused more so on the movement's challenge to the legitimacy of the Indian and Jharkhandi states as a self-defeating move. On the one hand, these particular interviewees affirmed the right of adivasis to control their traditional lands with little to none of the hesitation displayed by their nonprofit counterparts. On the other hand, they could not bring themselves to condone what they perceived as the movement's unnecessarily bellicose methods, averring that Pathalgadi agitators should contrarily secure their rights through the "proper channels"—presumably, by petitioning state authorities to protect adivasi territories from corporate appropriation, by removing the Das regime from power during the next election, and, at most, by staging peaceful protests as opposed to militant direct action to solicit support for their cause.

None of these activists harbor any illusions about the threats posed to Indigenous and other oppressed populations by the previous Das state government or the incumbent Modi national government. They certainly recognize how the much-vaunted institutions of Indian liberal democracy have been captured by proto-fascist forces, not least of all because they have been harassed, arrested, and threatened with reprisals for their solidarity with the most oppressed sectors of Jharkhandi and Indian society. And yet—perhaps out of fear for their hyper-vulnerable compatriots, perhaps due to their ideological and ethical commitments, but certainly with the best of intentions—they have maintained their faith in the processes that are meant to undergird these institutions, if not the current opaque, duplicitous, and unaccountable iterations of these institutions. This faith, which stems partly from the class backgrounds of the activists and intellectuals in focus, put them at odds with the Pathalgadi Movement.

The aforementioned faith in liberal institutionalism shone through even when I returned to Jharkhand in 2019 and found that many of these activists and intellectuals had actively and courageously engaged the Pathalgadi Movement in the previous year. Regardless of whether they still questioned the movement's weaponization of adivasi tradition, they had rushed to the defense of the thousands of participants arrested by the Das government prior to the latter's defeat in the 2019 Legislative Assembly Election. This laudable show of support was nevertheless couched in the language of liberal democracy. It was a challenge to the government's contravention of participants' civil liberties and a call for dialogue between state authorities and participants, not a wholehearted endorsement of the movement's political project (The Wire Staff, 2019). In keeping with this general stance, respected adivasi sociologist Virginius Xaxa (2019) contends that Pathalgadi Movement participants were "merely asserting the rights provided to them by law and the Constitution" in the face of the state's ignorance of these rights, even if they went "somewhat overboard with regard to the interpretation of some of the provisions." Eva Davidsdottir (2021) goes further in insisting that the Pathalgadi Movement operated "within the letter of the law." She explicitly frames nonviolent formal-legal struggle as legitimate and its violent counterpoint, in turn, as illegitimate, which once again begs the disconcerting question of condoning state violence against more militant movement participants. The discourse of liberal democracy may well have been strategically necessary to enlarge the movement's base of support and help its participants navigate the less than hospitable corridors of judicial power. However, it reaffirms the persistent tension between civil society and oppressed populations that are by and large excluded from its domain. Actors within these distinct political domains can clearly collaborate in productive ways, but the limitations of middle-class liberalism and its alienation of many grassroots interlocutors could prevent longer term coalition-building.

The frequent class divide between civil society actors and the communities they engage is highly pertinent to prospects for solidarity because nonprofit organizations and middle-class activists, intellectuals, and journalists are often the first points of contact for outsiders who wish to work with the communities in focus: they translate the realities of these communities into terms that they believe these outsiders will understand. This can put scholar-activists seeking solidarity in a complicated position, as they may well be unfamiliar with the sociopolitical landscapes they have to navigate, and their outsider status might prohibit their entry into communal spaces, particularly when these spaces are under attack by the state and capital. Scholar-activists can't afford to presume that their middle-class interlocutors speak for the communities they represent in their entirety. Long-term participant observation can allow scholar-activists to step beyond the boundaries of civil society and accompany militant social mobilizations. In the meantime, they would do well to recognize and monitor class dynamics, among other categories of differentiation, within the communities they engage and the larger social, political, economic, and cultural systems in which they are situated. Failure to do so can make acts of solidarity ineffective at best and counterrevolutionary at worst.

Thesis 3. Indigenous and other oppressed grassroots communities may foundationally challenge liberal democratic institutions at the same time as they strategically engage them, calling for a novel conceptual framework to understand these negotiations.

The Pathalgadi Movement's challenge to Jharkhandi civil society was part of its overall subversion of prevailing liberal democratic institutions—namely, the parliamentary, executive, judicial, and law enforcement mechanisms of the Jharkhandi and Indian states. Nevertheless, most of the movement's relatively few academic observers have situated its agitations along a liberal democratic spectrum, invoking grassroots or "popular" constitutionalism or, at most, legal pluralism (Sundar, 2018; Saxena and Chitkara, 2023). These invocations do not sit easily alongside the testimonies of several movement participants who condemned the failures of the Jharkhandi and Indian states to provide social services and uphold the promise of due process, questioning the very legitimacy of these governing apparatuses. One movement participant declared that "the judicial system only grinds people down—it does not deliver justice," while movement leaders repeatedly attested that their communities would not participate in elections or national occasions like India's Republic and Independence Days. A group of adivasi youths who gathered in Khunti's Arki block in April 2018 expressed even more explicitly anti-statist sentiments: "We do not recognize the Central or State governments or the President, Prime Minister or Governor. Our *gram sabha* [traditional adivasi village council] is the real constitutional body." Another group of young leaders echoed these sentiments: "Pathalgadi are basically a way to demarcate our territories and tell outsiders (government officials) that the law of the land does not apply here" (Tewary, 2018).

In more concrete terms, many Pathalgadi communities actually rejected their governmental entitlements to food, rural employment, housing, and pensions. Decrying the pilferage that plagues such schemes, villagers claimed that they could survive without them because they had been doing exactly that for years. "Out of the budget of Rs. 12,000 for a *Swachh Bharat* [Clean India] toilet," complained movement participant Budhua Munda to journalist Priya Ranjan Sahu, "government officials and contractors eat up more than a half . . . What is the point of such development" (Sahu, 2018)? Notwithstanding other movement participants who grounded their beliefs and practices in the Indian Constitution and pledged their non-opposition to the Indian nation, the Pathalgadi Movement clearly contained a strong current of autonomy. Nevertheless, this autonomy could only ever be contingent and thus incomplete as long as the state and capital persist, which forced the Pathalgadi Movement to make strategic concessions to these hegemonic forces. However, these concessions did not automatically invalidate the emancipatory potential of the movement.

The movement demanded that all funds earmarked for tribal development be handed over to gram sabhas, which at first glance suggests a desire to cooperate with and thus implicitly recognize the legitimacy of state authorities. However, other aspects of the 11-point charter drawn up by the movement belie its request for tribal developmental funds. It demanded that the government stop sending adivasis to jail on the pretext that they are Maoists, all amendments to land laws permitting the sale of tribal land to non-tribals be scrapped, and all police and paramilitary forces be withdrawn from tribal areas (Tewary, 2018). This combination of sweeping demands for freedom from state occupation, surveillance, and persecution with a single petition for access to state resources may indicate that most of the Pathalgadi Movement's underside could have comprised a zone beyond the reach of the government, as opposed to seeking wholesale inclusion into the ambit of governmentality. This zone was nonetheless conjoined to a lone point of strategic engagement rendered inevitable for survival by the state's control over resources. Even this lone point of engagement—the communal demand for national and regional tribal development funds—by no means entailed good-faith partnership or even extended coordination with the state. It conversely comes across as an intermittent financial transaction merely meant to enable and complement direct control over land, forests, and water that deliberately circumvented block and district officials.

Given the deep-seated disenchantment with state governance and welfare distribution chronicled above, the node of the Pathalgadi Movement that engaged the state could also be described in terms of a dis/simulation intended to bypass civil society and leverage the state's obligations to entitlement-bearers. Though some operated under the banner of the Constitution, many Pathalgadi participants seemed to have little interest in being governed. The constitutionalist face that they maintained for strategic ends appears to have camouflaged a zone of subaltern existence detached altogether from the state and civil society that actively eschewed and refused participation therein. This zone of incomplete autonomy was neither parallel or adjacent to nor compatible with its hegemonic counterparts, nor did it seek to replace them: it simply existed outside of and deliberately apart from them to the fullest extent possible.

Limited, highly strategic engagement with the state reappropriates resources for the survival of subaltern actors while still keeping the state at a safe distance. Dis/simulation reinforces this survival strategy by articulating the mobilization at hand in terms that are simultaneously familiar and unfamiliar enough to inhibit counterinsurgency measures. And disengagement from the state fosters the cultivation of alternate, autonomous lifeways. Together, these maneuvers frame a *repertoire of ungovernability*, wherein the subaltern actors at hand do not force their governing authorities to learn *how* they would prefer to be governed, but rather how much they would prefer to *not* be governed, as exemplified by their deliberate self-obfuscation and subsequent complication of their categorization and management.[3] This repertoire entails a multidimensional political strategy that requires all three maneuvers, though subaltern actors do not necessarily employ all three at the same time or all the time.

Scholar-activists seeking solidarity with Indigenous and other oppressed grassroots communities are all too often beholden to what I term *epistemic statism*, which I understand as a deep-seated orientation toward the state as a guarantor of societal democracy, welfare, and security at the expense of other forms and spaces of political action. The Pathalgadi Movement's repertoire of ungovernability brings this orientation into question: it shows how the neoliberal

---

[3] I draw upon Diana Taylor's (2020) definition of repertoires as embodied practices—such as dances, sports, rituals, or, for that matter, certain forms of collective mobilization and organization—that circumvent erasure by the archival strategies of colonialism, imperialism, capitalism, and statecraft.

authoritarian capture of the liberal democratic state—which is increasingly evident in many parts of the globe—drives subaltern communities to engage in politics by other means, with their remaining entanglements in statist mechanisms by no means signifying consent to sovereignty, governmentality, and coercion. It begs the question of how far scholar-activists are willing to go for the sake of solidarity: will they uphold liberal democratic institutions and risk alienating or even endangering their interlocutors, or will they accompany their subaltern coresearchers into the turbulent terrain of ungovernability, at potential risk to themselves?

Redefining Solidarity for an Era of Mounting Crises

The Pathalgadi Movement's contingent repertoire of ungovernability was by no means invulnerable to the state's monopoly on legitimate violence. The mass arrests suffered by the movement at the hands of the Das government severely impacted its reach and strength. Media coverage of the movement dwindled to sporadic reports providing technical updates on the legal proceedings involving key arrestees. And then, in February 2021, a group of about 100 adivasis from the Gumla District came to Ranchi and tried to install a large stone plaque in front of the Jharkhand High Court. They claimed that the land on which the Court sits is Indigenous territory where the government has no executive powers and that the recently elected government is unconstitutional. When questioned, they insisted that their actions had nothing to do with Pathalgadi whatsoever (Times of India, 2021). Furthermore, in March 2021, social workers in five villages in Khunti attempted to convene a two-day-long meeting to discuss how the 1996 PESA Act, a national law guaranteeing adivasi control over traditional lands, resources, and cultures, should be implemented in their areas. Due to police pressure, the convenors were forced to cancel the event, and they too disassociated themselves from Pathalgadi (Saran, 2021).

Is the Pathalgadi Movement experiencing a resurgence in all but name? If it is, however, will the new Pathalgadi Movement be as misunderstood and misrepresented as its predecessor, as it is once again refracted through the lens of liberalism by its opponents, its observers, and even some of its outside supporters?

As neoliberal Hindu nationalism has tightened its hold on Indian politics since the election of Narendra Modi in 2014, liberal constitutionalism, legalism, and welfarism seem to have taken on a far more subversive and even revolutionary hue among the country's progressive and left-leaning scholars, activists, and political commentators. Unfortunately, as this chapter has shown, these optics are inadequate to understand and appreciate many of the subaltern political formations that have emerged to contest ethnonationalism, neoliberal capitalist exploitation, extractivism, displacement, and state surveillance and repression in Jharkhand and beyond.

In dialogue with environmental historians Felix Padel and Vinita Damodaran, renowned Jharkhandi adivasi human rights activist Gladson Dungdung clarifies that present-day adivasi movements—including the Pathalgadi Movement— are answering a rallying cry shared by tribal icon Birsa Munda in 1900: *"Ulgulan ka ant nahi hoga"* ("Revolution will not end") (2022b: 1663). If this is in fact the case—and the Pathalgadi Movement very much suggests that it is—scholar-activists seeking solidarity with Indigenous and other oppressed grassroots communities must be prepared to examine not only negotiations in the state's long shadow but also political escape from this penumbra altogether. This analytical, political, and ethical imperative is perhaps even more crucial for personal, organizational, and material solidarity with subaltern actors. The latter may well be alienated by appeals from intellectuals and activists to collaborate and coordinate with state authorities, NGOs, news media, and other agents of bourgeois-oriented civil society. The Pathalgadi Movement's repertoire of ungovernability neither called upon elected representatives or judicial mediators to resolve its issues. It did not seek entry into civil society for the sake of accessing civic services and welfare schemes. Like so many other subaltern frameworks and tactics emerging across India and the Global South, it contrarily invites a careful rethinking of all these fixtures of liberal governance, illuminating the rich, dynamic political spaces, strategies, and collectivities thriving beyond its grasp.

References

Davidsdottir, Eva. 2021. "Our Rights Are Carved in Stone: The Case of the Pathalgadi Movement in Simdega, Jharkhand." *The International Journal of Human Rights* 25(7): 1111–125.

Dungdung, Gladson. 2022a. "The Pathalgadi Movement for Adivasi Autonomy." *Countercurrents*, March 1. Retrieved November 24, 2022. https://countercurrents.org /2022/03/the-pathalgadi-movement-for-adivas

Dungdung, Gladson. 2022b. "And Red Flows the Koina River: Adivasi Resistance to the 'Loot' of Their Land and Resources in Eastern India, 1980–2020." *Modern Asian Studies* 56: 1642–71.

Majumdar, Rakhi. 2017. "Jharkhand Attracts over Rs. 3.1 Lakh Crore worth of Investment." *The Economic Times*, February 17. Retrieved November 24, 2022. https://economictimes.indiatimes.com/news/economy/finance/jharkhand-attracts -over-rs-3-1-lakh-crore-worth-of-investment/articleshow/57209707.cms?from=mdr

Minority Rights Group International. 2022. "Adivasis." London, England: Minority Rights Group International. Retrieved November 24, 2022. https://minorityrights .org/minorities/adivasis-2/

Sahu, Priya R. 2018. "The Constitution Set in Stone: Adivasis in Jharkhand are Using an Old Tradition as a Novel Protest." *Scroll.in*, May 14. Retrieved November 24, 2022. https://amp.scroll.in/article/878468/the-constitution-set-in-stone-adivasis-in -jharkhand-are-using-an-old-tradition-as-a-novel-protest

Saran, Bedanti. 2021. "Pathalgadi Movement Gets More Refined, Demands Implementation of Rules." *The Hindustan Times*, March 19. Retrieved November 24, 2022. https://www.hindustantimes.com/cities/ranchi-news/pathalgadi-movement -gets-more-refined-demands-implementation-of-rules-101616172925508.html

Saxena, Astha, and Radhika Chitkara. 2023. "Decolonizing Sovereignty and Reimagining Autonomy: Adivasi Assertions and Interpretations of Law." In V. Clave-Mercier and M. Wuth (eds.), *Decolonizing Political Concepts*, 133–52. Oxford: Routledge.

Shah, Alpa. 2010. *In the Shadows of the State: Indigenous Politics, Environmentalism, and Insurgency in Jharkhand, India*. Durham, NC: Duke University Press.

Shahdeo, Kunal. 2022. "Will Adivasis be Able to Stave Off Attempts to Saffronize Them?" *Outlook*, November 24. Retrieved November 24, 2022. https://www .outlookindia.com/national/will-adivasis-be-able-to-stave-off-attempts-to-saffronise -them--magazine-239473

Singh, Anjana. 2019. "Many Faces of the Pathalgadi Movement in Jharkhand." *Economic and Political Weekly* 53(45): 28–33.

Singh, Sunil K. 2018. *Inside Jharkhand*. Ranchi: Crown Publications.

Sundar, Nandini. 2018. "Pathalgadi is Nothing but Constitutional Messianism, so Why is the BJP Afraid of It?" *The Wire*, May 16. Retrieved November 24, 2022. https:// thewire.in/rights/pathalgadi-is-nothing-but-constitutional-messianism-so-why-is -the-bjp-afraid-of-it.%20Accessed%2014%20August%202018

Taylor, Diana. 2020. *¡Presente! The Politics of Presence*. Durham, NC: Duke University Press.

Tewary, Amarnath. 2018. "The Pathalgadi Rebellion." *The Hindu*, April 14. Retrieved November 24, 2022. https://www.thehindu.com/todays-paper/tp-opinion/the -pathalgadi-rebellion/article23532618.ece

Times of India. 2021. "Jharkhand: 'Pathalgadi' Movement Resurfaces." *The Times of India*, February 24. Retrieved November 24, 2022. https://timesofindia.indiatimes

.com/city/ranchi/after-cms-tribals-not-hindu-remark-pathalgadi-movement
-resurfaces/articleshowprint/81178307.cms

The Wire Staff. 2019. "Activists Protest Repression of Tribals Engaged in Pathalgadi
Movement." *The Wire*, July 22. Retrieved November 24, 2022. https://thewire.in/
rights/pathalgadi-movement-jharkhand-tribals

Xaxa, Virginius. 2018. "Coercive 'Development.'" *Economic and Political Weekly* 53(45):
7–8.

Xaxa, Virginius. 2019. "Is the Pathalgadi Movement in Tribal Areas Anti-
constitutional?" *Economic & Political Weekly* 54(1). Retrieved May 27, 2024. https://
www.epw.in/journal/2019/1/alternative-standpoint/pathalgadi-movement-tribal
-areas-anti.html

2

Challenging Hegemonic Water Governance

Ontological Plurality in Sikkim, India

Dawa Yangi Sherpa

[1]Hydropower dams or hydroelectric projects (HEPs) are rooted in colonialist structures that establish economic dominance by converting free-flowing rivers into profitable commodities. Globally, the increase in energy consumption due to industrialization and urbanization has reframed rivers as commodities for electricity generation. The modern states in the Global South have embraced the "hydraulic mission" to strengthen state legitimacy, further contributing to water grabs. In response, global anti-dam and water (in)justice movements have challenged hydropower dam projects as they have displaced millions of people, mostly from Indigenous communities. Buttressed by institutionalized state power, these water grabs disrupt the ecological, Indigenous, cultural, social, and economic river systems; threaten water sovereignty; and lead to territorial and transboundary river struggles. The academic literature and civil society have considerably examined the socioeconomic impacts of hydropower dams. However, there has been limited analysis of epistemologies that integrate biodiversity and culture to challenge the dominant ontology of modern states' water governance structures. Therefore, emphasizing ontological differences between Indigenous and colonial as well as orthodox development perspectives on water could be a pathway to promoting ontological plurality in water

---

[1] This chapter builds on my MSc. thesis at The New School, New York, where I conducted semi-structured interviews (2020–2022) to explore the socio-economic impacts of hydropower dams. Shifting from the typical ecological destruction narrative, the chapter highlights local cultures, worldviews, and traditional stories, drawing from my time in the region (2008–2010). I am deeply grateful to everyone in Sikkim who generously shared their time, traditional stories and knowledge. Special thanks to Dr. Kalzang Dorjee Bhutia, Dr. Rinchu Doma Dukpa, Dr. Charisma K. Lepcha, and Dr. Mona Chettri for their work, which has illuminated Sikkim's complex lived realities in Himalayan Studies and Environmental Humanities. I also extend my gratitude to Dr. Amy Holmes-Tagchungdarpa and Dr. Kalzang Dorjee Bhutia for their invaluable feedback on local terminologies.

governance. In particular, engaging with Himalayan Indigenous worldviews, land-based cultures, and environmental ethics could be an alternative to global environmental discourses which are primarily dominated by exclusionary processes in international climate policy.

Since the early 1990s, the Indian state, with its "nation-building" agenda, has proposed a myriad of HEPs throughout the eastern Himalayas. Sikkim, an ecologically fragile eastern Himalayan state with colonial history, is undergoing postcolonial geopolitical reconfigurations of its physical and cultural landscape. Sikkim[2] lies within the Himalayan Massif, a transnational area with a constellation of multiple identities stitched together and divided by geopolitical demarcation (Shneiderman, 2010: 311). The introduction of HEPs in the region has exacerbated the interethnic and intra-ethnic local tensions in a region where the political landscape has adhered to party-based political clientelism and ethnic identity-based politics (Chettri and McDuie-Ra, 2018). In addition, the entry of internationally financed projects expedited HEPs in Indian states largely situated along the Himalayan region, including the Northeast borderlands of Sikkim. The resulting trajectories of water injustice coexist with other challenges proliferated by water commodification, resulting in water depletion, water deprivation, water insecurity, and, in the case of HEPs, permanent alteration of ecosystems.

In May 2003, when the prime minister of India launched the 50,000 MW Indian Hydroelectric Initiative as an economic development plan, it triggered a "hydro-rush" in the eastern Himalayas. This economic development agenda of the state has separated biodiversity from local land-based cultures and adversely impacted the local communities' place-based belief systems and traditional relationship with their sacred landscape. In response to the ecologically destructive hydropower dams, the local opposition to hydropower dams in Sikkim has opened spaces to embrace ontological plurality by respecting the existence of multiple deities, spirits, or divine manifestations in water governance. For several decades, the local ethnic communities have also coalesced through mutual understanding of respecting water and land-based relationships across the state of Sikkim. Simultaneously, the local ethnic communities have engaged in alternative forms of sustainable economies which contrast the state-sponsored extractive hydropower industries. In the process, the complex, deep-rooted heterogeneous ethnic identities of Sikkim claimed political agency

---

[2] Sikkim is a multiethnic, multilingual state with traditional communities and religious pluralism that go beyond those mentioned in this chapter which explicitly focuses on Sikkim's Buddhist worldviews and Rongkup (Lepcha) cosmology.

by challenging India's central government and the Sikkim state government's legitimacy over state-sponsored hydropower dams. The local communities introduced spaces to engage with ontological plurality in water governance by taking agency of the narrative[3] informed by local cosmology and Indigenous worldviews. Particularly, the state-sponsored hydropower dams were seen as a threat to the survival of their communities, which reinvigorated cultural and Indigenous identities to defend territory.

Worldviews, Cosmologies, and Ritual Life of Sikkim

Sikkim has over twenty ethnic groups who are a part of a larger diaspora of the Himalayan region's ethnicities and cultures. This chapter will focus on some of the distinct cultural, territorial, and land-based communities, namely the Lepchas, Bhutias, Lachungpas, and Lachenpas of Sikkim's North and West districts. In particular, the chapter will center their worldviews and cosmologies as it relates to their land-based cultures' lived realities and knowledge systems, which will be discussed below in relation to their ongoing struggles surrounding hydropower dams in the region.

Dzongu: The Source of Rong's Origin and Life

The Lepchas, Indigenous to the eastern Himalayas, call themselves Rong, and hold Indigenous religious practices and beliefs such as bóngthíngism and munism or shamanism (Lepcha and Torri, 2016). Rong cosmology, transmitted orally, centers place-based elements like snow, water, and land surrounding Kongchen Kongchlo[4] (Mount Kanchenjunga) as the epicenter along with the rivers Rongnyoo (Teesta) and Rungnyit (Rangit) as two river spirits. Rongs believe that they are created from the snow of Kongchen Kongchlo. The traditional stories based on Rong cosmology narrate that Dzongu land[5] is "the source of Rong's origin and life." Therefore, Dzongu's landscape, in particular the

---

[3] As noted, the chapter explicitly focuses on worldviews and cosmologies while simultaneously recognizing the shift in local dynamics over the last two decades. This shift in ethnic politics, nationalism, and cultural politics of identity is discussed in the academic works of Dr. Mona Chettri, Duncan McDuie-Ra, Dr. Rinchu Doma Dukpa, and many other leading scholars of Northeast India.

[4] In this chapter, I will provide the widely recognized or administrative word of places or rivers after a locally recognized or locally understood word used in reference to a traditional story or cosmology. Thereafter, the locally recognized word will be used to center local narratives.

[5] Dzongu is an area within Sikkim (see map).

rivers, holds a cultural and ethno-political significance for all Rong in Sikkim and beyond.

First, for the Rong communities of Dzongu and beyond, the construction of hydrodams meant that hydroelectric companies would control the free-flowing Rongnyoo river. The disruption of the river flow would negatively affect Rong communities' geo-spiritual ties because they believe that after they die, their souls travel through the tributaries of Rongnyoo to one of Kongchen Kongchlo's five peaks of origin based on their clan. Second, the state government's approval of hydropower dam proposals in Dzongu challenged the future of customary ancestral lands and jeopardized core beliefs of Rong identity. Therefore, when the region became an intense ground for several protests, grassroots campaigns, and mobilization, the youth members of Rong communities surrounding Dzongu mobilized around the narrative to "Save Dzongu and defend Lepcha identity." Finally, the state-led development project frameworks, such as the Environmental Impact Assessment (EIA), fail to integrate the worldviews and cosmologies of local land-based communities. Therefore, following the approval of HEP proposals within the vicinity of Dzongu, youth members of the Rong community challenged the project proposals during public comment periods, citing concerns such as sociocultural, territorial, and temporal displacement. Thus, they opened space to assert their geo-spiritual identity based on Rong cosmology within the modern nation-state's water governance structures.

Sacredness of Beyul Demojong for Sikkim's Buddhist Communities

Sikkim's Buddhist cosmology considers Sikkim as Beyul Demojong or the Hidden Valley of Rice or abundance. Sikkimese Buddhist communities maintain peace and harmony with the transdimensional inhabitants residing in Sikkim's shared landscape through traditional practices such as a ritual based on a special prayer text called Denjong Nesol (Earth Propitiation) (Bhutia, 2021). They contend that if local transdimensional beings are not consulted or appeased through rituals, hardships, disasters, calamities, and fighting among the community members can result (Balikci, 2008; Bhutia, 2021). Water, in various forms such as sacred lakes, springs, rivers, and ponds, is an important element in Sikkim's Buddhist worldview (Erschbamer, 2021). The relationship with water can be observed in ceremonial rituals like Bumchu, an annual Buddhist festival of Sikkim which is held at the Tashiding monastery. During the ceremony, people can witness the sociocultural significance of the water brought to the monastery for blessings and divinations from Rathong Chu, a sacred river for Sikkim's Buddhist communities.

Similarly, hydropower dam construction was largely seen as a cultural threat to the existence of Sikkimese Buddhist communities (Balikci, 2008). Their primary concerns reflected lived realities informed by their Sikkimese Buddhist worldviews. They cited worries that the hydrodams would bring ecological destruction, displeasing the guardian deities of the land or local spirits that live in harmony with the people within Sikkim's landscapes. For example, in early 1995, Rathong Chu HEP was proposed within the vicinity of Demojong, one of the most sacred places for Sikkimese Buddhist communities. First, the hydrodam's proximity to Tashiding monastery, which holds the annual Buddhist festival, Bumchu, jeopardized the sanctity of Demojong. Second, the water for Bumchu, which is sourced from the sacred river Rathong Chu, would be polluted by hydropower dam construction. Purity of water for performing rituals is an important element in Sikkimese Buddhist festivals. Finally, the hydropower dam proposals were largely opposed by the Sikkimese Buddhist communities because maintaining the sacredness of Sikkim's landscape and their multispecies relationship through rituals are the basis of the Buddhist worldview. In the process, the grassroots mobilization created a space to uphold their religio-cultural practices based on Buddhist cosmology within the state's agenda for capitalist development.

Dzsuma: Self-Governing Indigenous Political System of Lachen and Lachung Valley

The valleys of Lachen (great pass) and Lachung (small pass) lie in the North District of Sikkim bordering Tibet (China) with two distinguishing characteristics. First, the region has a centuries-old traditional administrative system of the valleys known as Dzsuma. This system is composed exclusively of place affiliated members, namely the Lachungpas of Lachung valley and the Lachenpas of Lachen valley. The members hold the right to vote and elect administrative heads of their Dzumsas whose primary responsibilities entail maintaining cohesion within the community and organizing collective works for the benefit of the community (Bourdet-Sabatier, 2004). Second, for these communities, the mountains and glacial lakes are the abode of their Lhasung(s) (guardian deities), revered and feared in these high mountain knowledge systems (Dukpa et al., 2018).

In May 2003, when the state-led economic development plans initiated a hydropower rush surrounding the region, the communities exerted their Dzsumas to successfully reject all hydropower projects proposed between 2004 and 2010. For these communities, a primary concern was that allowing such destructive projects would threaten their territory-affiliated identity as Lachungapas and Lachenpas and the legitimacy of their Indigenous political systems. More importantly, as communities hold religio-cultural beliefs based on their high mountain knowledge systems, two additional factors were crucial in shifting toward a unanimous anti-dam position. The first was the belief that hydropower dams would adversely impact the sacredness of valleys. The second was the role of "vernacular statecraft," encompassing local practices and knowledge systems that shaped political and social life of Lachungapas and Lachenpas, which has successfully stopped all hydropower dams proposed in their valleys (Dukpa et al., 2018).

Cultivating Mutual Relationships through Coalition-Building

Sikkim's anti-dam coalition was contingent on historically and culturally favorable local conditions that initiated political action for reclaiming territories. In particular, the contemporary alliance formed between the Lhopo (Bhutia) and Rong (Lepcha), the pluralistic religious system formed by integration of Sikkimese Buddhist worldview and Rong cosmology, and the legal recognition of ancestral land rights for Lhopo and Rong communities became the cornerstone for cultivating mutual relationships to oppose state-sponsored hydropower dams.

Bhutia-Lepcha Alliance

The alliance formed between the Lhopo (Bhutia) and Rong (Lepcha), Bhutia-Lepcha (BL), extends from the thirteenth-century treaty of blood brotherhood alliance. This alliance was forged under the sacred mountain Kongchen Kongchlo (in Rong cosmology) or Kanchendzonga (in Buddhist cosmology). Every year, Pang Lhabsol, a major festival of Sikkim, is celebrated through rituals to commemorate the alliance and reinforce relationships with the sacred mountain deity. In addition, a common linkage bridging their ethnic identities is their land-based cosmologies that are tied to the five peaks of the Kongchen Kongchlo or Kanchendzonga, and Sikkim's landscape, including sacred rivers, caves, and lakes. The shared sacred space of identity informs their resistance narratives and coalition-building strategies to oppose hydropower dams proposed within their landscapes.

The Bhutia-Lepcha alliance forged several successful oppositions against proposed hydropower dams in Sikkim. For instance, the Government of India's 2003 proposal for a 50,000 MW Indian Hydroelectric Initiative in Sikkim led to an urgent meeting in Chungthang, where the BL members gathered to discuss the impacts of hydropower dams in their lands and concluded with a commitment to oppose the initiative. Similarly, during the peak of Sikkim's anti-dam movement in June 2007, the BL House in Gangtok was a primary location for holding an indefinite hunger strike. In addition, the BL youth mobilization against the Teesta V HEP in 2022 led to the signing of a Memorandum of Understanding (MoU), making it India's first MoU drafted by an intervention from a coalition of youth activists (Eden and Wangchuk, 2018: 70). In 2006, the BL youth mobilized against Panan HEP, eventually leading to the cancellation of the project in 2009. Apart from building a statewide movement to challenge the Indian state's legitimacy over the plans to build hydropower dams in the vicinity, sustained grassroots efforts like the indefinite hunger strikes and community convergence by BL members highlighted legitimate concerns over the hydropower dams' sustainability in an ecologically vulnerable landscape. Specifically, there were concerns about landslides, earthquakes, and flooding. Over the next few years, the sustained grassroots activism by the BL alliance, aided by investigative reports, led to the cancellation of four dams[6] out of the six dams proposed in Dzongu.

Notably, the dam movement in early 2000 based out of Dzongu was led by the "movement brokers," mostly educated youth, or "urban intelligentsia," who were exposed to global injustices of hydropower dams that have historically displaced and alienated Indigenous populations and who knew of the significant sociocultural impacts from large-scale infrastructure projects. As a generation experiencing drastic changes to their local understanding of their landscape due to modernization and (neo)colonial policies, the youth were well-situated to use state tools (public petition, advocacy campaigns, alliances, rallies, and hunger strikes) to protect the sacred. For example, unprecedented financial investment from international agencies, carbon markets, and private financial institutions was gaining traction in Sikkim. These financial incentives further influenced interstate and intrastate politics. In the process, the EIA led to public consultations in and around Dzongu. The successful public outreach had led to massive participation and discussion during the sessions giving leverage to

---

[6] Rukel HEP (90 MW), Ringpi HEP (160 MW), Lingza HEP (120 MW), and Rangyoung HEP (80 MW)

negotiate with the state. While it is necessary to engage with the state under their construct of water-as-a-resource, it is equally critical to recognize that the technocratic scope of state tools like EIAs is limited. Above all, the modern state has used EIAs to justify projects and compensate for development-induced loss of property. In the case of Sikkim, it established private property rights, facilitating the transfer of communal and ancestral lands.

Pluralistic Religious System

The pluralistic religious system formed by Buddhist and Rong cosmology coalesced to form a broader narrative, or "symbiotic indigeneity,"[7] of the sacred Byeul (Buddhist cosmology) integrated sacred landscapes from Rong cosmology. This established a local environmental ethics based on the religio-cultural significance of the water (rivers). Sikkim's Buddhism has integrated with Rongs' Indigenous religion, forming a pluralistic religious system which is widely practiced in Sikkim (Lepcha and Torri, 2016). For example, the BL alliance was strengthened in early 2008 when the Panan HEP was proposed which threatened the ecological and cultural sanctity of the Tholung Monastery, a historically and culturally significant sacred Buddhist site within Dzongu. While the resistance narrative may have been centered around the political identity of the Rong communities, it simultaneously emerged as a struggle to protect Buddhist religio-cultural practices. Thus, the pluralistic form of place-based belief systems elevated the significance of the rivers which are considered sacred in Sikkimese Buddhist cosmology and Rong cosmology.

Similarly, a central feature of the pluralistic religious system in Sikkim is the ritual life. The traditional and cultural practices of establishing relationships through daily rituals with the interdimensional species (local guardians and spirits) residing in Sikkim's landscape are indispensable for maintaining ties with the land. Particularly, for Sikkim's Buddhist communities, ecologically destructive projects would anger local deities or local spirits and transdimensional inhabitants of Sikkim's sacred landscape, resulting in hardships, disasters, calamities, and fighting among the community members (Balikci, 2008; Bhutia, 2021). For example, in response to Rathong Chu HEP, Sikkimese Buddhist communities performed several rituals as a response to hydrodams to appease

---

[7] A term noted by Erik de Maaker in his chapter "Reimagining spaces, species and societies in the Himalayas." It denotes Indigenous modes of facing non-Indigenous forces of change such as political divisions.

local deities and to stop the hydropower project proposals from HEP companies which they considered as "enemies of the land" (Balikci, 2008: 239).

Likewise, in northern Sikkim's Lachung and Lachen valleys, there was growing animosity and tension after the hydropower dam proposals were first announced in 2003. Ultimately, a ritual ceremony was held at a central monastery in 2010 where members of Lachung took an oath to never allow a hydropower company to enter their vicinity. In particular, the traditional systems of the valleys and the local cultural practice of shamanism (Chya)[8] were deployed. The lived reality of members of Dzumsa establishes modes of knowing, contrasting state tools like EIA, which has led to a unanimous anti-dam position (Dukpa et al., 2018).

Legal Recognition of Ancestral Land Rights

The legal recognition or entitlement of ancestral land rights of the Lhopo and Rong communities after the 1975 merger of Sikkim with India provides a legal framework to bring forth BL communities' land rights (Constitution of India, 1978). It also enables us to highlight the threat of hydropower dams to Lhopo and Rong cultural and Indigenous identities. However, land rights recognition has limited community participation due to unequal power structures shared by ethnic identities with the right to claim land (Dukpa et al., 2018).

In particular, for Rong communities, the primary concern was the hydrodam's sociocultural impact on Dzongu. The hydropower dam proposals subjected ancestral communal lands to land surveys, the formal allotment of land rights, and land privatization (McDuie-Ra, 2011: 86). Once land titles were allotted, the pro-development project planners could acquire land through legal frameworks, incorporating ancestral lands into a state or private territory. These project components led to access restrictions for community members in the region. For land-based communities such as Rong, access restrictions within their ancestral lands also threaten their cultural identity. Additionally, for the past several decades, the central Government of India has undermined local mobility in the remote northern regions of Sikkim due to the geopolitical sensitivity with neighboring regions of Tibet (China).

All in all, the Indian central government's hydro-colonialism in Sikkim reflects a microcosm of policies enacted by postcolonial states. However, the decades following the central government's announcement of launching the 50,000 MW

---

[8] The practice of "Chya" involves summoning local deities of the mountains to punish individuals who pose a potential threat.

Indian Hydroelectric Initiative in Sikkim led to waves of resistance, movement, and coalition-building against state-sponsored hydropower dams. Members of local communities have shared a collective vision to resist hydrodams based on their land-based cosmologies. In addition, while the underlying foundation for resistance is the revitalization of land-based identities, there are two interventions by local BL communities which have influenced the collective vision through cooperation, reciprocity, and mutuality. The first is practicing the local rituals based on Denjong Nesol through interspecies communication (Bhutia, 2021). The second is the role of "movement brokers" who have challenged the state and hydropower industries using tools like the EIA, MoU, litigation, grassroot campaigns, and advocacy (Chettri and McDuie-Ra, 2018; Wangchuk, 2007; Dukpa et al., 2018).

However, as is the case with most anti-dam movements led by civil societies across the world, Sikkim's mobilization has slowly shifted narratives to fit into the realm of environmental sustainability designed in response to neoliberal economic concerns. In this realm of global sustainability efforts, hydropower industries have adopted several safeguards to minimize negative environmental impacts. These include fish ladders or fishways to facilitate migration, plant trees to mitigate deforestation during the land clearing process, small hydropower dams (under 100 MW), and compensation to displaced communities. Nevertheless, these are false solutions centered on technocratic policies unbefitting to Sikkim's local land-based cultures, cosmologies, and multispecies relationships.

In particular, the concept of sustainability metrics based on electricity generation from hydropower dams is implausible. For example, Rathong Chu (30 MW), which would be classified as a small hydropower project, will clearly get approval under these sustainability metrics. However, the grassroots resistance led by Concerned Citizens of Sikkim (CCS) highlighted that the modes of viewing water (rivers) through a religio-cultural perspective are radically in opposition to technocratic sustainability frameworks set as benchmarks for mitigation in environmental policies. In addition, for an economy-driven modern state like India, leveraging environmental gains or sustainability is futile as economic development, or "nation-building," takes precedence as the central narrative in any given context. Therefore, it may not be as advantageous to rely on state tools like the EIA and Cost-Benefit Analysis (CBA) or to engage with sustainability metrics emphasized by government administrative authorities fixated primarily on economic gains.

Embracing Ontological Plurality in Water Governance

In Sikkim, the state-sponsored hydropower development agenda radically contradicts local environmental ethics based on Buddhist cosmology, Indigenous worldviews, place-based identity, and cultural beliefs. In 2003, the Government of Sikkim proposed the 50,000 MW Indian Hydroelectric Initiative in Sikkim threatening the religious sacrilege of the Lhopo communities. Additionally, it advanced private property rights systems by disrupting customary land tenure in the Rong communities of Dzongu. In the years to follow, a series of events exposed a coercive political culture (Dukpa et al., 2018) and political marginalization (Gergan and Curley, 2021: 2). This ignited community mobilization rooted in a place-based and territory-affiliated identity which has been the cornerstone of the opposition against the Governments of Sikkim and Governments of India.

Sikkim's anti-dam coalition was contingent on historically and culturally favorable local conditions that initiated political action by placing Indigenous conceptions of relatedness to territory to the forefront: 1) the contemporary alliance formed between the Lhopo and Rong which extends from their historical brotherhood alliance; 2) the polycentric religious system (Buddhism and Rong cosmology) which coalesced to form a "symbiotic indigeneity" that helped to establish local environmental ethics based on the religio-cultural significance of the water (rivers); and 3) the legal recognition of ancestral land rights for Lhopo and Rong communities after the 1975 merger of Sikkim with India became the cornerstone for cultivating mutual relationships to oppose the state-sponsored hydropower dams and their threat to these cultural and Indigenous identities. The complex, deep-rooted heterogeneous ethnic identities have introduced the potential of embracing ontological plurality in water governance by taking agency over the narrative informed by religious, cultural, and Indigenous worldviews. In the process, claiming political agency in the Himalayan Massif has transformed the hegemonic spaces to actively engage local communities in the decision-making process of the modern state which seeks to restrict cross-border environmental knowledge and practices.

While grassroots organizing has witnessed some success in negotiating with the state, it is heterogeneous and limited in participation due to unequal power structures shared by ethnic identities with the right to claim land, which are equally manifested in both pro-dam and anti-dam narratives (McDuie-Ra, 2011; Dukpa et al., 2018). In contrast, the resistance narrative frames the mobilization factors (ethnic, cultural, and religious preservation) amid centuries of political, ethnic, and religious history grounded on the academic knowledge produced by scholars, and simultaneously acknowledges the relevance of contemporary ethnic politics of Sikkim in the process of movement-building. Recently, Sikkim has had multiple waves of resistance to defend territories, thus giving rise to a new generation of

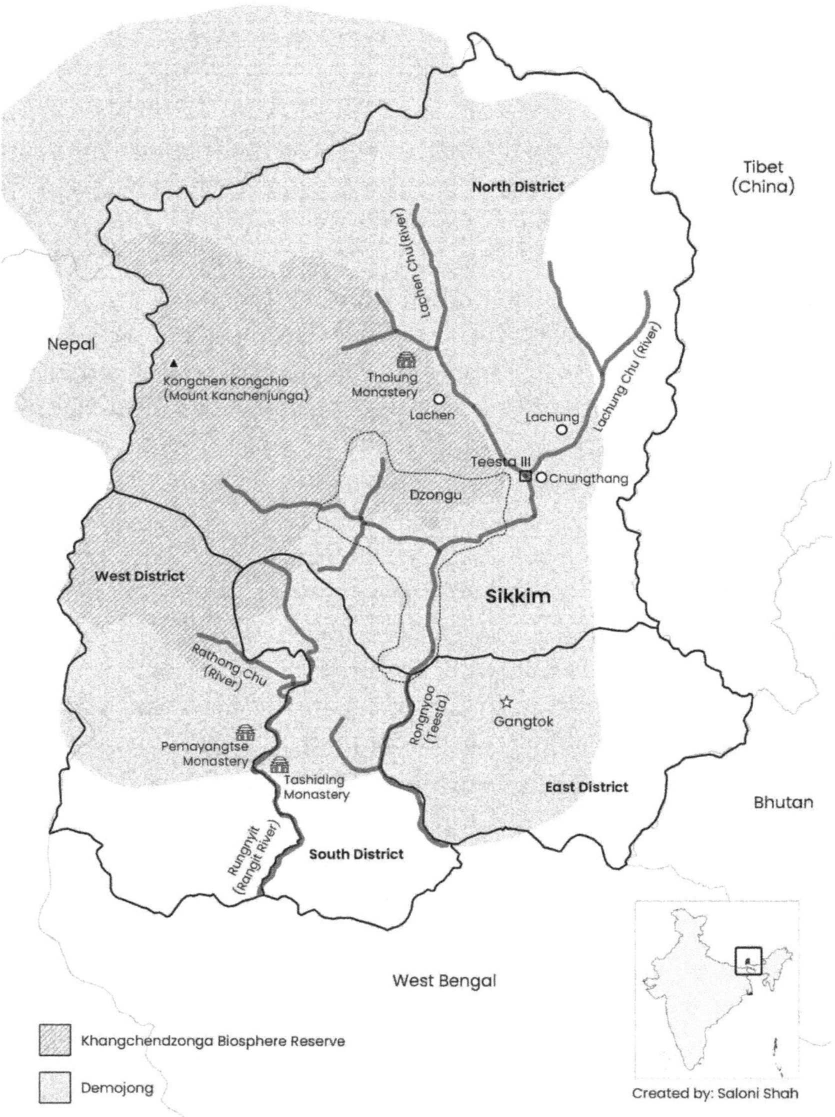

Figure 2.1 Map of state of Sikkim, India, with key locations noted in the chapter, *c.* 2023. Saloni Shah

defenders with decolonial narratives for the revalidation of Indigenous worldviews, customary law, and cultural ways of maintaining relationships with water (rivers).

The academic literature and both local and global civil society have engaged with the socioeconomic impacts of hydropower dams and influenced the ethno-politics surrounding hydropower dams in Sikkim. However, there has been little or limited engagement with epistemologies that bring together biodiversity and culture to destabilize the hegemonic ontology of water governance structures. Above all, these movements are dismissed and cited within the legal constitutional frameworks of the post-Independence Indian state as addressing religious issues or cultural constructs. Therefore, in Sikkim, the lived reality of local cultures and their ritual life based on Sikkimese Buddhist worldviews and Rong cosmology are discounted within the resistance narratives of Western, scientific academia and state water governance systems. In contrast, it is important to assert relational aspects highlighting ontological differences between Indigenous and colonial views of water to shift the narrative and create the potential of embracing ontological plurality in water governance. Similarly, within environmental humanities, the pluriverse or a "world of many worlds" of Himalayan Indigenous worldviews and environmental ethics is an emerging space with the potential to shift global environmental discourses which have been dominated by the exclusionary process of climate policy framing in international law.

References

Balikci, Anna. 2008. "Chapter Six. The Land, Its Problems and Ritual Solutions." In *Lamas, Shamans and Ancestors*, 216–34. Leiden: Brill.

Bhutia, Kalzang Dorjee. 2021. "Purifying Multispecies Relations in the Valley of Abundance: The Riwo Sangchö Ritual as Environmental History and Ethics in Sikkim." *MAVCOR Journal* 5(2): 10.22332/mav.ess.2021.1

Bourdet-Sabatier, Sophie. 2004. "The Dzumsa of Lachen: An Example of a Sikkimese Political Institution." *Bulletin of Tibetology* 93–104. http://www.thlib.org/static/reprints/bot/bot_2004_01_04.pdf

Chettri, Mona, and Duncan McDuie-Ra. 2018. "Delinquent Borderlands: Disorder and Exception in the Eastern Himalaya." *Journal of Borderlands Studies* 35(5): 709–23. https://doi.org/10.1080/08865655.2018.1452166

Constitution of India. 1978. "Article 371F: Special provisions with respect to the State of Sikkim." *Constitution of India.* https://www.constitutionofindia.net/articles/article-371f-special-provisions-with-respect-to-the-state-of-sikkim

Dukpa, Rinchu Doma, Deepa Joshi, and Rutgerd Boelens. 2018. "Hydropower Development and the Meaning of Place. Multi-Ethnic Hydropower Struggles in Sikkim, India." *Geoforum* 89: 60–72. https://doi.org/10.1016/j.geoforum.2018.01.006

Eden, Tshering, and Wangchuk Pema, Eds. 2018. *The Birds Have Lost Their Way: Essays on Hydropower and Climate Change Issues of Sikkim.* Gangtok: Lokta Books.

Erschbamer, Marlene. 2021. "Provider and Destroyer of Life: Legendary and Historical Buddhist Perspectives on Water and Its Socio-Cultural Significance in Sikkim." *Bulletin of Tibetology* 52(1): 59–74.

Gergan, Mabel Denzin, and Andrew Curley. 2021. "Indigenous Youth and Decolonial Futures: Energy and Environmentalism among the Diné in the Navajo Nation and the Lepchas of Sikkim, India." *Antipode* 55(3): 749–69. https://doi.org/10.1111/anti.12763

Lepcha, Charisma K., and Davide Torri. 2016. "Fieldwork in Dzongu: In Siiger's Footsteps and Beyond." In Ulrik Høj Johnsen, Armin W. Geertz, Svend Castenfeldt and Peter B. Andersen (eds.), *In the footsteps of Halfdan Siiger*, 147–62. Moesgaard Museum.

McDuie-Ra, Duncan. 2011. "The Dilemmas of Pro-Development Actors: Viewing State–Ethnic Minority Relations and Intra-Ethnic Dynamics through Contentious Development Projects." *Asian Ethnicity* 12(1): 77–100. https://doi.org/10.1080/14631369.2011.538220

Shneiderman, Sara. 2010. "Are the Central Himalayas in Zomia? Some Scholarly and Political Considerations across Time and Space." *Journal of Global History* 5(2): 289–312. https://doi.org/10.1017/s1740022810000094

Wangchuk, Pema. 2007. "Lepchas and their Hydel Protest." *Bulletin of Tibetology* 43(12): 33–57.

3

Scoping Coercive Technologies of Extraction

Lessons from Struggles against Land Grabbing in Tanzania

Felix Mantz

Land grabbing continues to be a central site of struggle, especially for communities in the Global South. Land grabbing is intimately tied to global environmental crises as well as the erosion of peasant and Indigenous peoples' power and autonomy over land and forms of social, political, and economic organization that can offer solutions and pathways for global socio-ecological liberation (Figueroa Helland et al., this volume; Sherpa, this volume). Land grabbing and resource extraction often require armed violence and direct coercion (Dunlap, 2020). Examining struggles against land grabbing in Tanzania, this chapter identifies a multitude of such "hard" coercive technologies and institutions mobilized by the state and capital to advance land grabbing and suppress resistance. These complement "soft" strategies that center on convincing communities to acquiesce to extraction and land grabbing. Where communities do not comply, "hard" coercive technologies and institutions directly expose movements, activists, and resistant communities to the state's monopoly of violence. Hence, movement "success" often depends on their ability to respond to and protect themselves from such efforts in order to defend against land grabbing. Scholar-activists and organizers who seek genuine and active solidarity with struggles against land grabbing should thus seriously consider how they can enhance communities' defensive capabilities.

This chapter's analysis of coercive technologies and institutions that use direct violence to advance land grabbing in Tanzania unfolds in three sections. First, I discuss the connections between land grabbing and global environmental crises.

Second, I provide a brief overview of land grabbing in Tanzania and scope the state of resistance against land dispossession. In the third part, I examine different "hard" coercive institutions that are mobilized to enforce land grabbing and suppress resistance. I then end with a short reflection. This chapter is based on reports, secondary literature, and research I conducted between 2018 and 2022. This research involved document analysis, archival research, and sixty-two interviews with Indigenous, farmers, and land rights activists, members of international conservation and development agencies, and representatives of Tanzanian state institutions. This chapter should be of particular interest to scholar-activists, civil society members, and those who (intend to) work with or in support of grassroots movements and who are concerned about how to engage in active and genuine solidarity.

Land Grabbing and Global Environmental Crises

Many of the chapters in this volume examine resistance to some type of dispossession of land and its resources—from rivers in the Himalaya (Sherpa) to minerals in Jharkhand (Raghu), grasslands in East Africa (Nangiria and Lunch), and forests in Cambodia (Nou) and Liberia (Kaba and Sillah). As all these chapters highlight, land grabbing is a complex and multifaceted process. It involves the alienation of power over territories from communities who live on (and partly off) the land by dominant entities operating within local, regional, and global political and economic contexts. Land grabbing is intimately tied to contemporary global environmental crises in several ways (Borras et al., 2012, 2020; Weldemichel, 2021).

First, the present realities and dire prospects of climate change and ecosystem collapse have animated states, transnational corporations (TNCs), investment funds, and wealthy individuals to grab control over land to secure access to coveted "natural resources" (Sulle and Nelson, 2009; Borras et al., 2012, 2020). For instance, the global food crises and the threat posed by climate change to the global food system result in scrambles for fertile soils, arable lands, and freshwater sources. Further, the reliance of the global energy system on finite hydrocarbons has intensified the search for other sources of fuel, such as biofuels, hydropower, solar, and wind, all of which require control and access to significant tracts of land or water (Sulle and Nelson, 2009; Borras et al., 2012, 2020).

Second, land grabbing contributes to the intensification of ecological and climate crises. In the context of a capitalist world-system that requires the consumption of more and more materials (so-called "natural resources") for its unceasing expansion and the accumulation of profit, controlling territories, lands, and ecologies is imperative (Frame, 2023). Put differently, land grabbing is one of the strategies employed by what Alexander Dunlap and Jostein Jakobsen (2020: 7) call the *Worldeater*: the "amalgamation of violent technologies and spirit propelling the global capitalist economy toward total extractivism." Total extractivism entails the consumption of all life through the systemic extraction of minerals, metals, fuels, water, cash crops, plants, and animal species which violently reconfigures Earth. This requires some form of power, control, and access over land that has to be grabbed—expropriated, accumulated, or seized.

Third, the climate and environmental crises have been used by powerful entities to justify the grabbing of lands and eviction of communities for conservation, wildlife protection, carbon schemes and other offsetting projects, and ecotourism (Benjaminsen and Bryceson, 2012; Oakland Institute, 2018, 2021, 2022; Bergius et al., 2020). These forms of land grabbing mostly operate on a Cartesian ideology, aiming to separate the "Human" from "Nature." This means that they mostly ignore the fact that humans have cared for, curated, and lived together with landscapes, biodiversity, wildlife, and ecosystem for millennia (Figueroa Helland et al., 2021). The "ecosystem services" and aesthetic value of specific landscapes which conservationists want to preserve and ostensibly protect from humans are often the results of complex forms of Indigenous land tenure, stewardship, and socio-ecological management (Homewood and Rodgers, 1991; Goldman, 2011; Figueroa Helland et al., 2021).

Fourth, land grabbing can emanate from profound desires for development, progress, advancement, and civilization, often understood in terms of emulating the Global North (Figueroa Helland and Lindgren, 2016). Pursuing these desires through industrialization, modernization, and large infrastructure projects entails the domination, disciplining, and control over lands and ecosystems by both state and corporate entities—what Adelman (2015) and others call the "mastery of nature." Not only is this form of development a central reason for contemporary climate and environmental crises, but it also reproduces a deep-seated anthropocentrism that must be disposed of in the search for other, more just worlds and futures (Figueroa Helland and Lindgren, 2016).

Land Grabbing and Resistance in Tanzania

Resistance to land grabbing must therefore be a central political project for any vision of global socio-ecological liberation. In Tanzania, land is dispossessed for several reasons. First, a large part of land grabbing is for the production of cash crops such as sisal, coffee, cashews, and others through large-scale modernized agriculture (Sulle, 2015; Bluwstein et al., 2018; Bergius et al., 2020; Frame, 2023). There are also hundreds of thousands of hectares allocated for biofuel production, especially jatropha, sugar cane, and palm oil (Sulle and Nelson, 2009; Sulle, 2015). Second, land is grabbed for conservation purposes and tourism (Sulle, 2015; Bluwstein et al., 2018; Bergius et al., 2020; Frame, 2023). Today, approximately 40 percent of Tanzania's mainland is designated as some type of conservation area. To differing degrees, all of these conservation areas enable the state and investors to erode people's power over land (Benjaminsen and Bryceson, 2012; Weldemichel, 2021). Third, lands are grabbed for speculative purposes and mining (Sulle, 2015). Mining constitutes Tanzania's main export sector and attracts most foreign direct investment (FDIs) (Bluwstein et al., 2018; Frame, 2023). The Tanzanian government claims that 90 percent of its land mass has potential mining resources. There are an estimated 480 mines in Tanzania, 125 of which are gold mines. Approximately 140,000 km^2 of land (15 percent of Tanzania's territory) has been designated for prospect licenses (Bluwstein et al., 2018).

Contemporary resistance against land grabbing in Tanzania has several goals and envisions different futures that vary depending on the specific context, including its communities, histories, organizers, and ecologies. These goals and futures are complex and at times contradictory. While some strive for inclusion, recognition, and rights, a significant number of struggles aim for different forms of self-determination, and autonomy (from the state, markets, and/or dominant models of development). Not infrequently, autonomy and recognition are pursued at the same time. Those elements of the struggle striving for autonomy want to make their own decisions over what happens to their lands, what type of development should be pursued, and how ecologies ought to be engaged with and related to.

Presently, there is a range of strategies employed to respond to and resist land grabbing. Within Tanzania, communities and allied organizations have sought to protest and lobby politicians to halt land grabs or negotiate better terms for land deals and to affect policy change in favor of smallholders. In addition, much of the organizing involves mobilizing rural communities, raising awareness of land rights, and in some instances, advancing legal cases to challenge land grabs. Everyday forms of resistance, refusal, and disobedience are employed as well. This includes, for instance, "trespassing" and otherwise continuing to use and access lands that have been grabbed. Another key strategy has been international organizing through, for instance, the Indigenous People's Movement, La Via Campesina, and various media campaigns, including participatory video (Nangiria and Lunch, this volume). This has drawn global attention to land grabbing in Tanzania and is likely preventing even more aggressive attempts at displacement and dispossession.

One struggle against land grabbing in Tanzania that has been especially successful in capturing international attention is that of the Maasai against the expansion of various conservation schemes. This struggle starts at the very beginning of German colonialization in the late nineteenth century and carried on throughout the British and then postcolonial and neoliberal periods in Tanzania (Neumann, 1995; Oakland Institute, 2018). Conservation regimes have drastically expanded throughout these periods with devastating consequences for Maasai and others, undermining their ability to survive as an Indigenous people. Indeed, the current state of territorial autonomy and self-determination for the Maasai is dire as efforts to grab their lands continue to advance (Oakland Institute, 2018, 2021; Mantz, 2023). Nonetheless, the Maasai are relatively well-organized and have developed high-profile strategies against land grabbing. In addition to forms of negotiation, community organizing, and building international networks, they have employed more confrontational tactics such as, sabotaging land grabs (for instance, by uprooting beacons used by the state to demarcate land) and engaging in civil disobedience. In doing so, they regularly encounter forms of direct violence from security forces organized by the state and capital. The following section discusses three "hard" coercive technologies and institutions that produce and legitimize such violence: borders, police, and prisons.

Coercive Technologies of Conservation Regimes in Tanzania

Borders, police, and prisons are vital institutions employed to expand territorial control by capital and state actors in the pursuit of land grabbing for conservation. They are also central to counterinsurgent strategies against resistance to land dispossession. In Tanzania, much of land grabbing for conservation takes place in areas that have been historically used by pastoralists and builds on deep colonial histories of displacing and eliminating pastoralists to create a mythical image of untouched African wilderness (Nelson, 2003; Neumann, 1995; Oakland Institute, 2018, 2021, 2022; Shivji, 2009). For example, in northern Tanzania, conservation areas have been established by removing Maasai communities in particular (Neumann, 1995; Oakland Institute, 2018, 2021, 2022).

Borders

Border practices are central to Tanzania's conservation regime. They keep people out of specific areas, assert state control, and reproduce an image of pristine nature and untouched wilderness that can be sold to conservationists and tourists alike (Neumann, 1995; Nelson, 2003; Goldman, 2011; Oakland Institute, 2018, 2021, 2022). In addition to evictions and displacement, borders are key to reconceptualize people, especially pastoralists, as trespassers on their own land. As a result, the movement of mobile people becomes criminalized and pastoralists themselves become criminals within conservation territories (Goldman, 2011; Oakland Institute, 2018, 2021). A similar strategy of criminalization is also applied to organizers who mobilize against land grabbing. Several reports as well as interviews I conducted highlight that authorities at times claim that such organizers are Kenyans rather than Tanzanians (Oakland Institute, 2018). One pastoral activist testifies that "if you start speaking, they will . . . [claim] the migration issue: either you are a migrant from Kenya . . . that you are not a Tanzanian, you are illegally here . . . or they can even peal the case of hunting,"[1] claiming that activists are involved in illegal hunting within conservation areas. Indeed, members of international conservation organizations as well as state agencies constantly referred to the problem of poachers harming wildlife as a type of criminality that needs to be addressed, including via "law enforcement." Hence, the activist explains, "people suffer a lot. You find most of the people, even those who can speak now, they don't have their voices heard because they are threatened"; they think "it is better now to keep quiet because our life is in danger."[2]

These claims of fighting illegal migrants and poachers are intended to portray those resisting land grabbing as outsiders, criminals, and as not belonging to the land. This also helps to undermine the legitimacy of resistance. Further, it

---

[1] Interview (November 21, 2020).
[2] Interview (November 21, 2020).

continues colonial discourses of identifying mobile people as threats to ecologies and wildlife. By being criminalized, pastoralists are policed, fined, subjected to overt violence, incarcerated for crossing conservation borders, and frequently have their cattle confiscated (Oakland Institute, 2018, 2021, 2022). As one representative of the Tanzania National Park Authority (TANAPA) admits, "we do arrest [and fine] a lot" because they "bring livestock within the Parks."[3] He elaborates that "conservation does not need politics or democracy" and that thus "the best thing is to stand firm and implement the laws." A member of the German Frankfurt Zoological Society (FZS), which has been involved in land grabbing for conservation since the early years of German colonization, also stressed the importance of coercive technologies and direct force: "I do believe law enforcement sits at the core of conservation . . . It's just as we don't leave our prime jewels or our gold bulletins on the street, you know; they are put in safe places, and they are protected. That's the same for wildlife and wild places too."[4]

Border regimes for conservation in Tanzania, according to several activists, are becoming increasingly militarized. This expands the state's capacity to surveil communities, enforce land grabbing, and employ direct violence to quell resistance and disobedience. For instance, in reference to the Ngorongoro Conservation Area Authority (NCAA), which polices the Ngorongoro Conservation Area (NCA), one Tanzanian activist working with a Maasai development organization testifies that conservation is "being militarized," meaning that "there is a lot of brutality of the military, the game rangers and all that"[5] in and around NCA. A Maasai elder describes conservation schemes in Tanzania as "using bullets instead of wisdom"[6] while another vocal Maasai activist finds that in "almost every single conservation area now in Tanzania they are establishing a paramilitary." As a result, when "you cross [into conservation areas], they arrest you; [when] you cross, they beat you."[7] A member of the Tanzanian Wildlife Authority (TAWA) admits this when explaining that TAWA rangers' "equipment is military equipment with the guns and firearms, and they are trained like a military . . . they are called a paramilitary group."[8] As a result, a high-profile Tanzanian opposition politician concludes that "[i]f you go around

---

[3] Interview (March 16, 2021).
[4] Interview (November 9, 2020).
[5] Interview (November 14, 2020).
[6] Interview (November 30, 2020).
[7] Interview (October 3, 2020).
[8] Interview (February 25, 2021).

the National Parks everywhere in Tanzania today, it is a story of murder and mayhem."[9]

Police

Police are central to enforcing the bordering practices of Tanzania's conservation regime in multiple ways. First, as described, police forces patrol protected areas in attempts to prevent people from entering. This includes state-based conservation institutions that become militarized and turn into policing agents, such as TAWA and TANAPA. These militarized conservation agencies are funded not only by the Tanzanian state but also rely significantly on transnational conservation groups and NGOs who provide monetary donations, equipment, and data. For example, according to several activists, the FZS has gifted and outfitted TAWA with equipment such as trucks that are used to police conservation areas (see also Mhagama, 2022). Similarly, a member of TANAPA reports in reference to FZS that "they have been funding vehicles."[10] He adds that the African Wildlife Foundation (AWF) has "been funding development of our plans," including the "training of rangers and other field officers—let's call it capacity development in general."

Police forces also demarcate and implement conservation borders. An example of this are the evictions of the Maasai from Loliondo in the summer of 2022. Loliondo is an area in northern Tanzania, bordering Kenya to the north and the Serengeti National Park to the west. In the summer of 2022, government officials planned to upgrade Loliondo from a Game Controlled Area to a Game Reserve (Amnesty International, 2022; FPP, 2022; UNHR, 2022). This upgrade has devastating effects on Loliondo's inhabitants, potentially displacing 70,000 Maasai (UNHR, 2022). These evictions are ongoing, with some Maasai having already been removed (Lee, 2023). Prior to this round of dispossession, during the 2017 evictions, police forces and park rangers also burned *bomas*, traditional Maasai homes, on government orders (Oakland Institute, 2018).

In 2022, to enforce this upgrade to a Game Reserve, approximately 700 police and military personnel, park rangers, and other security forces arrived unannounced in the territory and informed the Maasai that they would raise beacons to demarcate the new Game Reserve (FPP, 2022; UNHR, 2022). After community members uprooted the beacons and camped on-site in protest, the

---

[9] Interview (February 23, 2021).
[10] Interview (March 16, 2021).

police started shooting teargas and live ammunition at the pastoralists. Thirty-one people were injured, and one person was killed (Amnesty International, 2022; FPP, 2022; UNHR, 2022). This initiated ongoing confrontations between police and Loliondo's inhabitants that include police violence through imprisonment, torture, injuring and killing of inhabitants, raids, property seizure, land dispossession, and forcing thousands of people to flee, including across the border into Kenya (Amnesty International, 2022; FPP, 2022; UNHR, 2022; IWGIA, 2022). Hence, one pastoralist asserts that "the Maasai . . . are afraid of the police most of the time." He adds that an

> officer of law is not the protector of the Maasai people . . . they are our enemies, because most of the time when we see them, it's for taking us [away]. When we see them [it] is to put somebody in custody. Whenever we see a police officer, it was never for good intentions.[11]

Police are also mobilized to target pastoralist and Indigenous organizations in efforts to disable them and their operations, and thus to suppress resistance. For example, according to some activists, following a 2018 report from Oakland Institute, Tanzanian police believed they identified who the report's key informants were (see also Lang, 2018). They then started to shut down specific organizations by cutting off funding and shrinking their ability to operate. During my research, activists reported that they are regularly harassed and followed. For example, a member of an allied organization in the Global North reports that one especially prominent leader of the pastoral women's rights movement "has really received a lot of threats. She has been imprisoned and interrogated and her passport has been taken away from her for many years. And she's been accused of being a foreigner, a Kenyan."[12] Speaking of the same person, another pastoralist from Loliondo adds that "even one of her bothers was killed once in Thomson [Thomson Safaris—a infamous US-based tourism company operating in Loliondo] area."[13]

Prisons

For several years, the Tanzanian government has been using prisons to quell regime critique and dissenting voices (Amnesty International, 2019). Based on some activists' accounts, prisons have become increasingly important

---

[11] Interview (November 21, 2020).
[12] Interview (October 20, 2020).
[13] Interview (November 21, 2020).

in Tanzania to buttress the bordering and policing of conservation areas. Specifically, activists or their friends and family are regularly arrested. As Dunlap (2020) explains, imprisonment or the threat of detention has several functions in the context of land grabs. First, prisons are used to intimidate and bully activists to stop their work. Second, the prison constitutes a tool of deterrence. By arresting activists and leaders, their constituent communities can be deterred from organizing against land grabbing. As a result, other, less confrontational responses to land grabbing such as selling lands or cooperating with authorities and NGOs become more attractive. Third, imprisoning key people who resist land grabbing at key moments undermines their movement-building work. It also robs communities of important advocates and people who occupy central roles in the social reproduction of the communities themselves. This can be observed in the Loliondo evictions of 2022. After state representatives announced the reclassification of Loliondo, but before the evictions took place, several elders were arrested and locked up at unknown locations (FPP, 2022; UNHR, 2022). Only after these elders were removed did the police come in, erected the beacons, and then attacked those pastoralists who resisted. The leaders were released a few days later (FPP, 2022; UNHR, 2022).

Employed in this way, prisons also become institutions that reproduce the racialized criminalization of the Maasai and assert the legitimacy of state presence and violence. As leading abolitionist Ruth Wilson Gilmore (2022: 274) explains, because "prisons and prisoners are part of the structure of the state, they enable governments to establish state legitimacy through a claim to provide social 'protection' combined with their monopoly on the delegation of violence." She elaborates that "[t]he state establishes legitimacy precisely *because* it violently dominates certain people and thereby defines them (and makes them visible to others) as the sort of people who should be pushed around" (Gilmore, 2022: 274). Applied to the Tanzanian context, this suggests a symbiotic relationship between the use of coercive technologies (including borders, police, and prisons), state and conservationist claims of criminality in and around conservation areas (e.g., accusations that those engaging in resistance are illegal migrants, trespassers, poachers, and lawbreakers), and desires to further expand conservation regimes via land grabbing.

In sum, much of the organizing against land grabbing in Tanzania must contend with coercive institutions such as borders, police, and prisons that expose communities to direct forms of violence in the state and capital's attempt to gain power over lands. This reality suggests that a key question which those who seek to be in active solidarity with struggles against land grabbing must ask is how they can enhance frontline communities' ability to defend against and deter the use of direct violence. Connectedly, how do we weaken the capabilities of state and capital to unleash coercive technologies of direct violence? These questions boil down to the necessity of urgently redressing the state's monopoly of violence. They also directly engage what Samir Amin identifies as capital's monopoly over military force, for which, as Samuel Grant highlights, "we [still] lack sufficient answers" (Grant, this volume). For starters, I suggest that Indigenous struggles for territorial autonomy and against settler colonial desires for land, as well as abolitionist organizing and strategizing against police, prisons, and borders, constitute rich archives of tools, tactics, and experiences that we should engage to inform transnational solidarity with struggles against land grabbing in Tanzania and beyond.

The Carrot and the Stick

Coercive institutions and technologies that unleash direct violence to advance land grabbing and suppress resistance are rarely employed in isolation. Instead, they are complemented by "soft" political techniques of persuasion, enchantment, and manipulation (Dunlap, 2020). These aim at pacification by convincing people that land grabs and conservation projects are in their favor. According to Dunlap (2020: 673), "'soft' counterinsurgency approaches seek to gain 'social license' or win 'hearts' and 'minds'. This is accomplished by employing—real or imagined—positive mechanisms to obtain legitimacy in a self-reinforcing and economically sustainable way." One example is promises of social development, whereby state and corporate actors try to convince people that they will benefit from tourism or receive income from setting certain areas aside for conservation. Promises of the empowering effects of land titles also fall into this category. Several activists I talked to explained that villagers are pushed to pursue individual land titles with the promise that they gain land tenure security and can access bank loans for entrepreneurial ventures. However, land titles can benefit the state and TNCs seeking control and access over land (Mantz, 2023). Specifically, land titles have led to the individualization and commodification of rural land as well as the erosion of communal land tenure systems that can protect against land grabbing and make it more difficult for the state and capital to control territories. Put simply, having land owned individually rather than communally means that it can be more easily acquired by TNCs, especially when rural dwellers are poor and more tempted to give up their land cheaply to access much-needed cash in the short term (Mantz, 2023).

Another example of soft techniques that complement coercive technologies is the mobilization of "scientific" knowledge in support of land grabbing for conservation. Several activists pointed out that organizations that have been involved in Tanzania's conservation regime, such as UNESCO, IUCN, and FZS, continue to promote ideas about wildlife and conservation that are used to legitimize land grabbing (see also Gardner, 2016; Oakland Institute, 2021, 2022). For example, Global North conservationists whom I spoke to (affiliated with UNESCO, FZS, and IUCN) explicitly or implicitly blame population and population growth as the key threat to wildlife and biodiversity (see also Oakland Institute, 2021, 2022). These sentiments were echoed by representatives of Tanzanian state-based conservation institutions who work closely with these organizations. As mentioned previously, this ignores the fact that people can (and especially Indigenous peoples often do) contribute to biodiversity and ecosystem health, including in densely populated areas. What is most important is the prevailing social, cosmological, and material relations between humans and the web of life within a particular land base.

Conclusion

A key lesson from examining land grabbing in Tanzania and resistance to such land grabbing is that when communities and movements refuse to acquiesce to the territorial desires of capital and the state, and instead enter into struggles for autonomy, land defense, and self-determination, they will likely face a multitude of coercive technologies and institutions. Complementing "soft" techniques of dispossession, these expose movements to direct violence and force. Thus, the success of sustained land defense, movements against land grabbing, and struggles for territorial autonomy seems to at least partly depend on whether they can continuously and effectively respond to the multiple technologies and institutions of direct violence advanced by state and corporate entities. Hence, an important task for allied organizations, activists, scholars, and members of civil society who seek to be in active solidarity with land defense and movements against extractivism and socio-ecological crises is to think about ways to enhance their capacity to effectively face and deter these technologies of direct violence, and to disable the state and capital's ability to wield them.

References

Adelman, Sam. 2015 "Epistemologies of Mastery." In Anna Grear and Louis J. Kotze (eds.), *Research Handbook on Human Rights and the Environment*, 9–27. Northampton: Edward Elgar Publishing.

Amnesty International. 2019. *The Price We Pay: Targeted for Dissent by the Tanzanian State*. London: Amnesty International.

Amnesty International. 2022. *Amnesty International Report 2021/22: The State of the World's Human Rights*. London: Amnesty International.

Benjaminsen, Tor A., and Ian Bryceson. 2012. "Conservation, Green/Blue Grabbing and Accumulation by Dispossession in Tanzania." *The Journal of Peasant Studies* 39(2): 335–55.

Bergius, Mikael, Tor A. Benjaminsen, Faustin Maganga, and Halvard Buhaug. 2020. "Green Economy, Degradation Narratives, and Land-Use Conflicts in Tanzania." *World Development* 129: 104850.

Bluwstein, Jevgeniy, Jens Friis Lund, Kelly Askew, Howard Stein, Christine Noe, Rie Odgaard, Faustin Maganga, and Linda Engström. 2018. "Between Dependence and Deprivation: the Interlocking Nature of Land Alienation in Tanzania." *Journal of Agrarian Change* 18(4): 1–25.

Borras, Saturnino Jr., Elyse Mills, Philip Seufert, Stephan Backes, Daniel Fyfe, Roman Herre, and Laura Michéle. 2020. "Transnational Land Investment Web: Land Grabs, TNCs, and the Challenge of Global Governance." *Globalizations* 17(4): 608–28.

Borras, Saturnino Jr., Jennifer C. Franco, Sergio Gómez, Cristóbal Kay, and Max Spoor. 2012. "Land Grabbing in Latin America and the Caribbean." *The Journal of Peasant Studies* 39(3–4): 845–72.

Dunlap, Alexander. 2020. "Wind, Coal, and Copper: The Politics of Land Grabbing, Counterinsurgency, and the Social Engineering of Extraction." *Globalizations* 17(4): 661–82.

Dunlap, Alexander, and Jostein Jakobsen. 2020. *The Violent Technologies of Extraction: Political Ecology, Critical Agrarian Studies and the Capitalist Worldeater*. Cham: Palgrave.

Figueroa Helland, Leonardo, and Tim Lindgren. 2016. "What Goes Around Comes Around: From the Coloniality of Power to the Crisis of Civilization." *Journal of World-Systems Research* 22(2): 430–62.

Figueroa Helland, Leonardo, Abigail Pérez Aguilera, and Felix Mantz. 2021. "Decolonize, Reindigenize: Planetary Crisis, Biocultural Diversity, Indigenous Resurgence and Land Rematriation." In Cathy Wagner and Luis I. Prácanos (eds.), *Contesting Extinctions: Decolonial and Regenerative Futures*, 23–62. Minneapolis: Lexington Books.

FPP. 2022. "Press Release: 'We Will Not Leave. Not Now. Not Ever': Tanzanian State Violently Seizes Maasai Land in Loliondo." *Forest Peoples Programme*, June 13.

https://www.forestpeoples.org/en/press-release/2022/press-release-maasai-land
-evictions-loliondo

Frame, Mariko. 2023. *Ecological Imperialism, Neoliberal Development, and the Capitalist World-System: Cases from Africa and Asia.* London: Routledge.

Gardner, Benjamin. 2016. *Selling the Serengeti: The Cultural Politics of Safari Tourism.* Athens: University of Georgia Press.

Gilmore, Ruth W. 2022. *Abolition Geography: Essays Towards Liberation.* London: Verso Books.

Goldman, Mara J. 2011. "Strangers in their own Land: Maasai and Wildlife Conservation in Northern Tanzania." *Conservation and Society* 9(1): 65–79.

Homewood, Katherine and W. A. Rodgers. 1991. *Maasailand Ecology: Pastoralist Development and Wildlife Conservation in Ngorongoro, Tanzania.* Cambridge: Cambridge University Press.

IWGIA. 2022. "Situation of Maasai in Loliondo and Sale in Tanzania Dangerous and Desperate." *IWGIA*, July 5. https://www.iwgia.org/en/news/4853-maasai-loliondo
-tanzania-desperate.html

Mantz, Felix. 2023. *Colonial Logics of Land Grabbing in Tanzania.* Doctoral thesis (Ph.D.), Queen Mary University of London.

Mhagama, Hilda. 2022. "Tanzania: Zoological Society Boosts Security Patrol in Selous Game Reserve." *AllAfrica*, March 10. https://allafrica.com/stories/202203100512.html

Lang, Chris. 2018. "Tanzania's Maasai Face Ongoing Threats, Intimidation, and Arrests." *Conservation Watch*, June 6. https://medium.com/conservationwatch/tanzanias
-maasai-face-ongoing-threats-intimidation-and-arrests-3e02d70c6858

Lee, Joseph. 2023. "Indigenous Maasai ask the United Nations to Intervene on Reported Human Rights Abuses." *Mongabay*, April 21. https://news.mongabay.com/2023/04/
indigenous-maasai-ask-the-united-nations-to-intervene-on-reported-human-rights
-abuses/

Nelson, Robert H. 2003. "Environmental Colonialism: 'Saving' Africa from Africans." *The Independent Review* 8(1): 65–86.

Neumann, Roderick. 1995. "Ways of Seeing Africa: Colonial Recasting of African Society and Landscape in Serengeti National Park." *Ecumene* 2(2): 149–69.

Oakland Institute. 2018. *Losing the Serengeti: The Maasai Land that was to Run Forever.* Oakland: Oakland Institute.

Oakland Institute. 2021. *The Looming Threat of Eviction: The Continued Displacement of the Maasai under the Guise of Conservation in Ngorongoro Conservation Area.* Oakland: Oakland Institute.

Oakland Institute. 2022. *Flawed Plans for the Relocation of the Maasai from the Ngorongoro Conservation Area.* Oakland: Oakland Institute.

Shivji, Issa G. 2009. *Where Is Uhuru? Reflections on the Struggle for Democracy in Africa.* Cape Town: Pambazuka Press.

Sulle, Emmanuel. 2015. "Land Grabbing and Agricultural Commercialization Duality: Insights from Tanzania's Transformation Agenda." *Afriche e orienti* 3: 109–28.

Sulle, Emmanuel, and Fred Nelson. 2009, *Biofuels, Land Access and Rural Livelihoods in Tanzania*. London: IIED.

UNHR. 2022. *Tanzania: UN Experts Warn of Escalating Violence Amidst Plans to Forcibly Evict Maasai From Ancestral Lands*. https://www.ohchr.org/en/press-releases /2022/06/tanzania-un-experts-warn-escalating-violence-amidst-plans-forcibly-evict

Weldemichel, Teklehaymanot G. 2021. "Making Land Grabbable: Stealthy Dispossessions by Conservation in Ngorongoro Conservation Area, Tanzania." *Environment and Planning E: Nature and Space* 5(4): 2052–72.

4

Participatory Video Is a Revolutionary Tool!

Samwel Nangiria and Nick Lunch

Indigenous peoples are some of the most affected by social and ecological crises today, yet they also hold the key to solving these crises. Four hundred years of resistance to colonialism and capitalism have given Indigenous peoples vital perspectives, and their voices in local, national, and global spaces must be amplified. However, they often lack access to communications technologies and training, rendering their stories and perspectives largely untold or at best ventriloquized by well-meaning outsiders.

This chapter shares the experiences and lessons learned from the Living Cultures Indigenous Fellowship and the Decolonising Cultural Spaces project. The fellowship is a groundbreaking strategy that delivers remote training to Indigenous peoples wishing to harness participatory media as a tool for engaging and mobilizing their communities. In this way, participatory video becomes a method used by Indigenous peoples to highlight the struggles and issues they face. The Decolonising Cultural Spaces Project seeks to address the process of marginalization that places the European experience at the center of historical narratives in museums. Objects, some of them sacred, are presented as artifacts of obsolete cultures. Incorrect information and ambiguous stories of "collection" are continued sources of pain for affected communities. The program supports Indigenous peoples as they address these deep-rooted historical problems. InsightShare works with communities as they spread the message that Indigenous cultures are living cultures.

This chapter takes the form of a dialogue that centers the testimonies and reflections of Samwel Nangiria, a Maasai activist and co-founder of the Pan African Living Culture Alliance (PALCA). The dialogue is facilitated by Nick Lunch, the co-founder and director of InsightShare, a participatory development organization working in support of Indigenous peoples' struggles for self-determination. In the first part of the dialogue, Samwel and Nick reflect on the need to decolonize media work and the role participatory video can play. In the second part, Samwel provides an account of the Maasai struggles for more balanced and just relations to land, highlighting how participatory video can be a vehicle for knowledge exchange, and a tool to build radical alliances and networks among Indigenous peoples themselves as well as with other allied and related struggles. In this context, Samwel then offers a reflection on efforts to decolonize museum spaces in the UK.

Decolonizing the Media and the Role of Participatory Video: A Conversation

Samwel: Capitalism has been the means and the route that colonialism has used to harm the planet. It has affected and disconnected human beings from nature, broken cultures, and destroyed sustainable lifeways. For the current generation to be able to survive and think about the next generations, capitalism has to be replaced, now and urgently. How to do it can't be the task of those who are responsible for the situation we are in. Indigenous wisdom and connections to Mother Nature provide urgently needed alternatives to save our planet.

Nick: There is a growing recognition of the urgent and central role Indigenous peoples might play in monitoring, mitigating, adapting to, and averting the worst possible outcomes of the climate crisis. But the majority of these communities lack the skills, platforms, and confidence to genuinely shape policy, solutions, and narratives locally, regionally, and internationally. When Indigenous communities stand up for their rights, they are challenged by an anti-Indigenous bias in national governments and neoliberal impositions of ecological imperialism.

Given how disproportionately at risk Indigenous peoples and their unique lifeways are from the climate shocks and biodiversity losses that have been caused by industrialized societies, we believe it is morally imperative that we all work faster to ensure these communities have the means to shape decisions and attitudes to nature, whether about their own lands or internationally.

We believe digital activism and participatory storytelling represent one of the most powerful and cost-effective ways of doing so. Equipping Indigenous groups with digital activism and participatory storytelling skills has been a potent means to reinforce cultures and spiritualities, promote intergenerational and intercultural learning, and build alliances.

Samwel: A group of community leaders from eleven Indigenous groups from across the breadth of Africa created the Pan African Living Cultures Alliance

(PALCA) in 2018. Our vision is for African Indigenous, pastoralist, fisher, and hunter-gatherer communities to be recognized and respected universally for their unique contribution to safeguarding the continent's biocultural diversity. PALCA connects our communities, strengthens resilience, and facilitates the sharing of traditional knowledge, technologies, and values for the well-being of all on the planet.

PALCA's mission is to achieve our vision through safeguarding communities' biocultural rights, supporting intergenerational transmission, and traditional governance of natural resources, with participatory video (PV) at its heart. We wish to see our communities continentally interconnected, able to secure biocultural resources and transferring traditional knowledge across generational frontiers.

I first came across PV through social media, watching videos made by communities like mine, and this sparked my interest. I read about InsightShare and their approach to skills development within Indigenous communities. At the time, I was organizing a campaign of mass awareness on the threats to my people's land rights. I had developed contacts with mainstream media in Tanzania and linked our elders with journalists for briefings on the sacred connections the Maasai have with their traditional homeland. Through this process, I came to know that Maasai elders and traditional leaders weren't comfortable sharing their innermost thoughts and feelings about their sacred link with the land.

I learned that mainstream media is not reliable, and most of what they publish or cover has to go through a chain of people with different interests. The news has to be edited and aligned with media house agendas. Most of the information shared by elders on land rights issues was never aired. Additionally, most journalists do not understand the complex relationship between Indigenous communities and their environment. I also learned that media is vulnerable! I have witnessed our news being blocked from airing, manipulated, or edited to focus on something that is not the people's priority. Having read about PV, how it is media made by and for Indigenous communities, how it is setting the record straight and enabling a revolution, I was inspired to be trained in its methods and become a PV facilitator. Essentially, I wanted to help our community challenge fake or modified news, particularly about our sacred struggle to protect our land and culture.

Nick: These efforts to set the record straight are important. At 5 percent of the world's population (300–370 million people), Indigenous peoples are custodians of over 80 percent of the Earth's biodiversity. With the Intergovernmental Panel on Climate Change (IPCC) declaring "code red" for our species, there is a huge surge in interest in Indigenous people's wisdom and stewardship of lands and ecosystems. Uniquely intertwined with their local environments, Indigenous peoples are particularly sensitive to environmental changes and vulnerable to environmental degradation. Indigenous lands are disproportionately at risk from extractive and polluting industries that accelerate climate change: mining, deforestation, and intensive land use.

With PV, Indigenous communities are able to address key issues that matter to them: protection of land, livelihoods, and culture. Through video documentation and community screenings, they maintain their ancestral wisdom and cultural diversity. PALCA is building a network to share knowledge and strengthen the power of Indigenous movements to shift public perspectives on Indigenous issues, peoples, traditions, and land. Over the course of twenty years, we have repeatedly witnessed the value and power of training Indigenous peoples in digital activism: participants develop the skills to champion and conserve their own communities' languages, culture, knowledge and lands, and often experience hugely positive and profound shifts in their personal sense of agency and optimism in the process.

In 2021, InsightShare piloted a seven-month leadership program for Indigenous women and youth to foster skills in facilitation, journalism, critical thinking, advocacy, digital activism and public speaking. Due to Covid-19, the course was delivered online and used familiar, secure, and low-bandwidth technologies—such as WhatsApp—to reduce costs and to leapfrog the lack of communications infrastructure in often-remote Indigenous areas. Thirty-eight Indigenous women and youth, between the ages of twenty and thirty-five from Kenya, Namibia, South Africa, and Tanzania, took part in the pilot program, creating community-authored videos on issues that mattered to them aimed at both local and global audiences. Each community set up an autonomous video hub to use as a training and production site, which together formed a regional network of media centers to enable Pan-African dialogue and exchange.

Samwel: I had to explain PV and its potential to our elders, our knowledge holders. After receiving training by InsightShare, I returned to the community with a camera. Elders told me, a machine like that is normally in the hands of white people! They believe that a camera is one of the many ways their images, cultures, and environment have been stolen. I informed them about the concept of PV and assured them that the camera and the equipment belong to them. It took some time to explain who gave me the camera and why and to assure them there wasn't anybody hiding behind me with interest to access their culture, their sacred stories. I also informed them that we will establish a group that will be filming and facilitating community screenings. After that, the elders accepted, and as a process of receiving, we had a blessing ceremony where elders blessed both the cameras and myself.

We formed a video collective and invited InsightShare to run a more in-depth workshop. After that we made the *Olosho Video*, which was screened to the Tanzanian parliament, garnering support for the community. Many women, elders, and youth participated in the interviews. We filmed all around the community. We screened the video back to the community who provided feedback to help us improve it. The making of the video helped to promote PV and built trust within the community. I continued using the camera, and we established *Oltoilo le Maa Tanzania* (Voice of the Maasai) to carry on the vision of decolonizing media through PV.

Nick: PV develops agency. Our Fellows' self-confidence grew radically: from low self-esteem to a profound belief in their capacity to affect change in their communities. Fellows learned the essential tools for successful participation in global forums and international webinars: technical skills, critical thinking, and the skills/tools for online collaboration. Fellows drew up plans to keep hubs operating after the life-span of the project, seeking to pass their skills on and explore storytelling as a tool for income generation.

The video hubs hold regular community screenings to improve the content of their films, thereby overcoming community skepticism of videos, film, and technology and forging new relationships of trust with elders. Our Fellows gain a strong understanding of broader issues impacting Indigenous communities through inter-hub exchanges and through the process of creating their films. Film topics include sacred sites, traditional education, food and nutrition, and tackling health and well-being issues like tuberculosis, teenage pregnancy, and gender-based violence. Fellows benefited from powerful online exchanges with other hubs in the fellowship: sharing their first videos, giving feedback, and discussing shared issues. Inter-hub dialogues increased participants' sensitivity to and appreciation for their own cultural uniqueness, provided inspiration for greater conservation of their own linguistic or cultural traditions, and revealed the corrosive impacts of Westernization on Indigenous identities.

Some video hubs in northern Kenya have identified storytelling as a tool to resolve and reduce the frequency of inter-tribal conflicts through fostering greater understanding and dialogue. For example, through screenings at cultural events, the Gabbra and Borana teams addressed the escalating violence from cattle raids in Marsabit. Our fellowship program is very effective in linking up video hubs around joint advocacy campaigns, from decolonizing conservation and challenging illegal land grabs to amplifying Indigenous voices on climate justice. Channel 4 News combined the video hubs' footage to broadcast a powerful five-minute cut on the opening night of UNCOP26, reaching global audiences of up to two million people with the voices of Africa's Indigenous women and youth!

The hubs built connections and pathways of support for one another, discovering shared struggles and designing partnership campaigns such as "Listening to the Land." This involved Maasai, KhoiSan, Xhosa, El Molo, Turkana, Rendile, and amaMpondo video teams from East and Southern Africa filming pilgrimages to sacred sites, forests, oceans, and local communities to speak with elders about climate change impacts. The films were screened during COP26 at prominent venues across Glasgow, including the Green Zone. Our work at COP26 and beyond aims to inspire people to reimagine development and conservation, and the way we relate to each other and Mother Earth.

From Colonial Conservation to Ecological Decolonization: Maasai Perspectives

Samwel: In the immediate situation in Loliondo and Ngorongoro (northern Tanzania), PV has been the source of real-time information, bringing international scrutiny to what is happening on the ground. Loliondo and Ngorongoro have been targets of conservation-motivated land grabbing since the colonial period starting in the late nineteenth century. The most disturbing video of security guards shooting at Maasai who protested the ongoing land grabs in Ololosokwan on June 10, 2022, was captured by a member of the Oltoilo le Maa video collective. The photos of the people who were injured were taken by the same members of the collective. The introduction of PV in Loliondo has made it possible to document new developments and quickly but surely connect people across the globe. The result has been an astonishing level of news coverage across the global media.

This century, we have had three other violent episodes in Loliondo—2009, 2013, and 2017. However, the current one of 2022 had the most coverage worldwide, which was possible due to PV's ability to communicate stories and new incidents. The Government of Tanzania has banned media from accessing Loliondo or even remotely covering the eviction. PV has strategically filled the gap, and through social media platforms, the stories were widely shared across the globe, resulting in widespread condemnation. With such video and photographic evidence of human rights violations in Loliondo, UN human rights bodies have been convinced to push for an independent commission of inquiry in Loliondo.

The Government of Tanzania has consistently denied any violation of human rights, denying the violent evictions and using conservation to justify annexing a further 1500 km² of our lands for elite tourism. The smartphones that we use for video documentation have the GPS and locations enabled. Because of this, our footage can be used in the courtroom as evidence to counter false accusations and lies. We have also interviewed the most affected people—those who were shot, human rights defenders, and those who had to run away in fear.

The Government of Tanzania, through President Samia Suluhu Hassan, had prepared for this eviction. The so-called "Royal Tour" was a preparatory event, whereby the President participated in a documentary that was frequently aired on national television and screened in different countries. Its purpose seems to be to raise the alarm for the need to protect the Ngorongoro reserve by evicting the Maasai from their homeland. The President used the Royal Tour to attack our culture, livelihood, and the connection we have with our environment. The film falsely asserts that the Maasai are "a primitive tribe" resisting progress and that we are "one of the newest" arrivals in Tanzania and not really Indigenous to the region. In our Community Report that we shared with the prime minister, we included both an analysis of the impact of the colonial conservation model in Ngorongoro and how our culture has resisted it. We also exposed the discrimination in national mainstream media. We have shown how the President's prejudices were in fact setting the ground for eviction. Through an in-depth comparative analysis, we have compared the health of wildlife and biodiversity in no-go zones and the areas where Maasai reside in the Ngorongoro Conservation Area. Our report has clearly displayed the lies behind the colonial narratives. It called for close re-examination of the dominant narratives about the conservation and protection of nature, which have enjoyed de facto acceptance within the existing frameworks.

Based on our experience, it is clear that PV is a timely weapon that Indigenous communities need for defending their human and cultural rights. The vision I have is for all Indigenous communities of Africa to be connected through PV and to have the skills to use it for defense. PV will enable them to articulate their historical and contemporary struggles and to collaborate and network within themselves and with outside partners.

Nick: The word "conservation"—what does it mean for you and how does it affect the lives of the Maasai people, especially those in Tanzania. What about other Indigenous peoples in other parts of the world? What is problematic about this word—the assumptions that lie behind it as being "a good thing"?

Samwel: Conservation to me means military, exploitation, business, and money. This word has been used to make people believe that nature is in an emergency situation that calls for rescue. Laws were made by the colonial masters to force humans and nature to separate. Through the dominant narratives and funding of conservation work, this perception has become entrenched. My community has been affected by the application of this colonial ideology. We have lost our homelands to the false claims of conservation. The government, with the support from global conservation lobby groups, attack us, disown us, and dispossesses us off our land by using laws enacted for this purpose or using the military to fight us directly. For example, conservation in Loliondo and Ngorongoro has always been associated with the income that it brings to the government. So, the more the income, the more effort the government puts into alienating us from that land. Our land has been taken and is still being taken with promises for development projects that aspire to improve our well-being. We are viewed and targeted by the government as people who are not patriots, who don't have love for their country, and who do not want to change. We are paying the price of our harmonious and balanced life with nature by shedding our blood and resisting their propaganda. Our economy has experienced sustained shocks, huge losses of livestock due to prolonged drought, and a lack of access to natural resources on top of violations of our human rights and the resulting poverty.

In other parts of the world, conservation has been used, just as in our case, to dislocate, evict people, destroy livelihoods, endanger cultures, and create poverty. From the Amazon to the Democratic Republic of the Congo, Indigenous and aboriginal people are being forced out of the forests to facilitate easy targets for logging, mining, and other extractive land practices that are profit-based. Conservation has never been good; it is as bad as colonialism itself!

After independence, the new government inherited the same ideology from the colonial rulers of demeaning our culture as primitive, disconnecting culture from land, insulting our traditional knowledge, and putting us into a sustained conflict to protect our lands, culture, and identity.

PV has been a handy tool. We have made many films, most of them revealing the way Maasai have been tirelessly taking care of their environment. Our first video, *Olosho*, and many others since have been inspiring people and attracting support through improving the understanding of the Maasai's role in conserving nature, not only globally but also within Tanzania. In the current crisis, we are experiencing large support across the globe, and this is connected to the role of PV.

Nick: What does decolonization mean to you? What does it mean to your community? Do you think the term is overused and what impact does this have?

Samwel: Decolonization to me means denying in the strongest possible terms the way of colonial thinking. It is about understanding what colonialism is all about, repairing the damages and holding to account the colonial systems, replacing them with people-centered mindsets. In my community, it means a continued struggle to block the colonial mindset and practices, despite the fact that such practices and mindsets have dominated across the world. The Maasai have always struggled to ensure their traditional land management, knowledge, leadership, language, and traditional dress remain intact within the changing sociocultural environment.

The term is overused to mean something else. The Maasai community feel governments, international development organizations, donors, the business community, and media refer to decolonization as a means of improving or modifying the dominant ideology but not replacing it. Development processes have been the tools to promote colonialism through a slogan of "change." Any process that is meant to deprive Indigenous and local communities of their resources, particularly lands, is normally blended with the idea of development, transformation, and change—all aiming at replacing what is existing.

We, as Maasai community, and of course the Indigenous communities worldwide, have resisted colonialism. We have stood firm to protect our lands and our cultures despite being isolated. The word belongs to the colonial masters. They are using it as an approach to gain relevancy in engaging communities in the Global South. Just like climate change, they have destroyed the planet and are now looking to our communities for the solutions, some of which, such as REDD+ and carbon trading, threaten our land security further.

Decolonizing Museum Spaces and the Work of the PALCA

Nick: Can you tell us more about PALCA, why did you set it up, what impacts are PALCA and the Living Cultures Fellowship achieving?

Samwel: PALCA is a movement of Indigenous communities to secure and retain their land security, cultures, languages, and a flourishing future. It was set up to create a safe space for us to remain in the drivers' seat, to promote self-determination as we navigate the challenges ahead.

     Grassroots Responses to Extractivism

For example, Laibon Mokompo, our Spiritual Leader, warned us there will be many people who will try to hijack the process of reconciliation and healing that we are starting with UK museums. There are those who will try to impose a different approach, such as the Maasai urban elites who were calling for legal action to sue the museums.

There are many others coming from outside our community, such as academics and other museums, who will come with different agendas and different priorities. For some institutions, this work may be a branding exercise to promote the role of museums and to talk about decolonizing because it is in fashion. But this is very dangerous. We must try to see who is really willing to engage deeply. So there must be a strong institution that is properly grounded in the community, which will act to protect the interests not only of our nation but also the integrity of this process and its outcome—to guide partners and ensure this is happening because it is simply being responsible, being open, and being honest.

So Mokompo was very willing to see something emerging in Africa, something from the Maasai community, something that brings Indigenous communities of Africa together to go beyond the Maasai and think about a united Africa. Mokompo was happy to see Indigenous communities across the continent, a circle of friends, who are struggling to keep their lands, to keep their knowledge, to keep their cultures, to remain relevant in this changing world. So he advised that PALCA be established to maintain the importance of living cultures at the heart of this work.

We have extensive video documentation of our latest trip to the Pitt Rivers Museum in Oxford in 2020 with the Maasai Spiritual Leader to reveal the stories of the problematic artifacts, how they were collected, and how they reached the museum. This enabled people to witness the quality of discussions, the kinds of questions we were asking, the radical way we were untangling the past. Combining our traditional knowledge system with the museum archives, we tried to know more about the artifacts. This made our people understand our role as representatives of the Maasai Nation: to recognize the museum's lack of knowledge of the artifacts and where we need to fit in as a community to ensure we control the narratives of what is on display. If it were not for PV, I'm sure this process could not have reached to where we are now because, for us, this is a very new process and people have a lot of questions, a lot of doubts. People didn't know where to start, how to engage in decolonizing the museum space, but given the support of PV, people are now learning more about how a museum looks like, how it functions. So, PV for me is quite an important tool to facilitate a discussion between two sides, where there may be a power imbalance. PV can bring some kind of leveling, bringing community people to the same level of sharing knowledge with academics and powerful institutions.

Nick: The idea of contemporary narratives from Indigenous peoples sits in tension with what museums are comfortable doing. By shifting the narrative about the past through these interventions, we create space for a different narrative about the present and the future. What about the current narratives of the Maasai of today, those which are missing from the museum? Do you think that PV could tell those stories? Should the museums include the stories of the present alongside the stories of the past?

Samwel: Through our work with the museum, the stories from these objects reveal a very close connection between what has happened to us and what is happening now. For example, the Maasai community has been struggling all along to try and manage the historical legacies perpetrated by governments as part of the way it delivers development programs.

PV is one of the best ways to reveal the current struggles, the stories of today, through the objects held by museums. Our stories make the artifacts collected even more precious! The concept of living cultures becomes something that is real. PV is one of the best ways, if not the only one, that museums can capture the current struggles and ensure that people are not only informed about the history but also about what is happening today, and how the past and the present are related. Perhaps that has not been the focus of the museum up until now, but with involvement and pressure from Indigenous peoples, they have no choice!

PV has become a link between the Maasai nation, museums, universities, partners like InsightShare, and many others who are willing to come in. PV has been able to support that link. This was a revelation to Mokompo! This is not just a technology, but it is something that is rooted in the knowledge of people, something that people trust, something that his community, his spiritual institution is putting a lot of trust in. He said the Maasai spiritual institution is willing to offer support and guidance from the spiritual point of view. PV has to be taken as an important process that not only links people but also builds capacity. So that needs spiritual guidance, because this is about engaging, about meeting, about refreshing historical issues, about crafting the future.

Conclusion

Nick: Please describe an alternative worldview provided by the Maasai community where you are from, and why is it relevant today?

Samwel: The Maasai worldview is about a balanced lifeway that doesn't claim superiority over other living and nonliving things. This includes a perception that is concerned about the future generations of Maasai and other things with whom we share the planet as our home. It also entails a sustainable people-centered natural resources management approach and leadership that is not based on interests—economic and political. With challenges of global warming and an uncertain future, the Maasai worldview could be an alternative to learn from.

My vision is to see my community being able to reclaim its autonomy in all aspects of life with their knowledge being at the center and respected. PV comes as an enabling bridge, documenting and sharing our worldview to inspire others, and therefore influence the process of decolonization.

Nick: Our field has grown rapidly in the past twenty years, and InsightShare is now one of a number of expert and agile organizations with a wealth of insight and experience in how to deliver this powerfully enabling work. However, much more needs to be done if we are to realize the opportunity of genuine Indigenous self-representation and for these groups to play their fullest part in ensuring our planet remains habitable at this perilous moment.

We believe now is the time to raise our collective ambitions and share our expertise, resources, and plans in order to rapidly increase the number of Indigenous groups accessing this training, to build a genuinely worldwide network of Indigenous media hubs, and to bring the skills of self-representation to as many groups as possible.

Our ambition is to engage with policymakers and governments, provide free advice, create multi-stakeholder platforms and collaborations that interweave scientific knowledge on climate change, practical policy-making and traditional Earth-based wisdom. We strive to establish international media, research, and advocacy partnerships to increase the reach and impact of films. We are scaling out our Indigenous-to-Indigenous learning model, by enabling current fellows to lead their own trainings in neighboring Indigenous communities. We continue to strengthen the capacity of Indigenous-led organizations by investing in local infrastructure, such as contributing to office and connectivity costs. We support advocacy and engagement by profiling our fellows in online and face-to-face events, exhibits, webinars, and conferences.

We have demonstrated some of the benefits to Indigenous peoples of digital activism and media hubs. But there is much more to do if we are to shift public perspectives on Indigenous issues, peoples, traditions, and land, and particularly to support Indigenous peoples to seize the opportunity to influence crucial policies that directly impact them, such as the implementation of global commitments to protect 30 percent of the world's land and ocean by 2030 through support for intact Indigenous territories and self-governance.

Building a network has the potential to hugely increase Indigenous movements' impact, effectiveness, resilience, and joint and individual lobbying power. This is why over the next five years, InsightShare aims to train and support 200+ Indigenous facilitators in twenty countries to bring video and storytelling skills to multiple Indigenous communities across Africa, Asia, and the Americas and we would like to work with like-minded individuals, communities, funders, NGOs, and research institutes to build the field of Indigenous digital activism.

5

Between the Forbidden Forests

Conservation in the Krahn-Bassa Proposed Protected Areas (KPPAs) in Liberia

Ali D. Kaba and Baba Sillah

This chapter focuses on local resistance against top-down forest conservation efforts in southeastern Liberia.[1] The country has an estimated 11 million hectares of land, almost 60 percent of which are forested, including 4.5 million hectares of tropical rainforest—the most extensive contiguous forest in West Africa. The forest is home to valuable wood species and significant biodiversity. Considering the extent to which anthropocentric activities have affected most ecosystems worldwide, Liberia has emerged as one of the focal points in global efforts to combat climate change.[2] As part of its fight against climate change, the country has protected at least 10 percent of its land and pledged to conserve 30 percent of its forests.[3] Meanwhile, Liberia is also desperately seeking foreign investments to spur economic development, with over half of its land area currently leased to companies undertaking topsoil and subsoil activities, including logging, agriculture plantation, and mining. Proponents of these acquisitions argue that they promote economic development and food security, but local communities and human rights groups complain that they often cause more harm than good, displace local communities, harm the environment, and exacerbate inequalities. They also undermine the country's conservation goals.

---

[1] This paper is based on field experience and interviews conducted in Monrovia and by telephone between July and August 2022. Specifically, we worked closely with James Otto of the Sustainable Development Institute (SDI), a Liberian nongovernmental organization engaged in land and resource rights advocacy.

[2] A revised Nationally Determined Contribution (NDC) in 2021 requires approximately $490 million to fund its green growth, mitigation, and adaptation strategies from 2020 to 2030.

[3] Liberia ratified the Convention on Biological Diversity and the National Biodiversity Strategy and Action Plan of 2004.

Paradoxically, Liberia is also undergoing land and natural resource management reforms, including the Forestry Reform Law of 2006, the Community Rights Law of 2009, and the Land Rights Law (LRL) of 2018. The laws grant communities the right to own and manage their communal lands and forest resources. However, in most cases, these legislations have not been properly implemented and are often left to the discretion of government agents and civil society organizations (CSOs). Meanwhile, Liberia's murky land history and poor policy implementation blur the line between law, practice, and custom (Steven, 2014; Alden-Wily, 2007). Until 2018, most of Liberia's land was held de facto by the state, leaving local communities as "squatters" without tenure security. Thus, large-scale land-use reforms, whether for commercial or conservation purposes, may negatively impact local communities, restricting access to natural resources and traditional land-use activities.

This chapter examines local resistance to the Krahn-Bassa Proposed Protected Area (KBPPA).[4] The KBPPA is a 290,167-hectare forest corridor in the southeast of Liberia and one of the largest national parks in the country.[5] Taking the KBPPA as a case study, the chapter illustrates how two communities, Senkwen and Juarzon, are resisting government-led conservation projects. The communities have employed diverse and overlapping strategies to protect their lands, resources, and livelihoods.

Conservation and Resistance against Land Dispossession

Over the past few decades, there has been a significant change in land management practices in rural areas across Africa. Top-down land management practices, environmentalism, and development discourse have all played a role in transferring rural land claims to public and private entities (Moyo et al., 2019). Reports indicate that almost 9 percent of the continent's arable land has been acquired by private investors since 2000 (Borras et al., 2012). Most of these land transfers occur in countries with poor governance, especially in postwar reconstruction countries like Liberia.

---

[4] National parks, protected areas, and conservation efforts are related but distinct terms. The terms will be used interchangeably to refer to areas that are legislated or regulated by the state to protect biodiversity, historical sites, or cultural objects.

[5] Community Ecoguard Patrol Handbook for Krahn-Bassa Proposed Protected Area. 2022. Critical Ecosystem Partnership Fund. Available at: https://www.cepf.net/sites/default/files/community -ecogard-handbook-kbppa-2022.pdf

While most land transfers have occurred between the state and private business enterprises, emerging discourse on climate change mitigation strategies has pushed for more protected areas worldwide. The International Union for Conservation of Nature (IUCN) defines protected areas as land areas—ecosystems and cultural heritage—that are protected, usually by some international-state-led design. Studies suggest that protected areas contribute to climate adaptation and prevent biodiversity loss (Dudley et al., 2017). A growing consensus among international policy institutions and environmentalist groups supports expanding protected areas. Protected areas currently cover 14.9 percent of the globe's surface and store approximately 12 percent of all atmospheric carbon dioxide (IUCN, 2019). Initial conservation efforts were perceived as a fortress under scientific regulations, a government imposition on people and space. Scholars have termed such conservation efforts "fortress conservation,"[6] contending it does not adhere to ethical or sustainable principles (Brockington et al. 2008; Adams, 2004).

Recent initiatives are moving away from the fortress approach to a framework that integrates local livelihoods into conservation efforts (Oldekop et al., 2016). The United Nations Educational, Scientific and Cultural Organization (UNESCO, 2008) recommends a balance between conserving biological diversity, promoting economic growth, and maintaining livelihoods and cultural values. Community involvement and integration into conservation efforts have recently been recognized as important conservation strategies. This new approach, which carries a narrative of community participation and comanagement regimes, has gained support from many international environmental groups, including UNESCO. Theoretically, integrating the community into the conservation process may result in individuals becoming environmental subjects (Agrawal, 2003). These conservation projects, while well-intentioned, often fail to consider the long-term impacts of their actions on local communities. Studies have revealed forced displacement, dispossession, and social disarticulation that threaten local livelihoods, land rights, and food security (Dressler et al., 2010). Furthermore, changes in land use and management systems significantly determine groups' and individuals' socioeconomic status, especially those who rely on land and resources for their livelihoods (Gilfoy, 2015).

In Liberia, activists claim conservation efforts are not discussed beforehand with local communities, which can cause frustration and resentment. In recent

---

[6] The term refers to a practice where protected areas are created and managed without considering the rights and needs of the local communities.

years, conservation efforts have often been met with resistance from local communities, ranging from peaceful protests to violent demonstrations. Such resistance is often a response to the perceived abuse of power and control by those leading the conservation efforts. The two main forms of community resistance are passive and active. Passive resistance is enacted by people who characteristically ignore conservation regulations and continue to do whatever they want. This form of resistance is often unorganized and individualistic (Scott, 1985). On the other hand, active resistance involves more organized and collective efforts to defy the rules. This type of resistance can take many forms, such as protests, sit-ins, or boycotts (Hollander and Einwohner, 2004; Scott, 1985; Holmes, 2007). Community resistance to conservation regulations can be shaped by various factors, including class, gender, generation, ethnicity, nationality, and historically specific expectations, aspirations, and traditions (Scott, 1985: 34).

A Brief Background of Liberia

Liberia is one of only two African countries that was never colonized by European powers. The country was founded in the early 1800s by the American Colonization Society (ACS), a philanthropic organization based in the United States of America to resettle freed slaves (Sawyer, 1992). The settlers employed a combination of treaties and threats to dispossess the Indigenous communities of their "tribal land," legitimized by laws and regulations (Stevens, 2014). This was achieved by institutionally transforming traditional leaders into state agents paid, regulated, and accountable to the state and local elites, which Catherine Boone (2014) has termed "neo-customary leaders." In many cases, these leaders hold dual roles as state agents and traditional leaders. Their dual roles often place them in a gray area of authority, where they hold substantial power over their constituents but limited local accountability, resulting in patron–client relationships with the state. Scholars have attributed Liberia's 14-year civil war (1989–2003) to this skewed power relationship (Unruh, 2009).

Since the Liberia War ended in 2003, the country has passed laws recognizing local communities' rights to their customary land and resources. One such law is the LRL of 2018, which protects communities' land and resources. However, despite the legal efforts, more than half of the country's land remains contracted to mining, logging, and agri-plantation companies on long-term contracts—see Table 5.1 (O'Mahoney, 2019). In addition, 30 percent of the country's forests

Table 5.1 Some Major Land Concessions in Liberia

Name	Hectares	Kind of Concession	Year
Sime Darby Plantation	330,000	Rubber	2009
Golden Veroleum Liberia	350,000	Rubber	2010
Firestone	405,000	Rubber	1926 (last renegotiated in 2008)
Forest management contracts	1,088,000	Logging and other resource extraction activities	

have been committed to international efforts at mitigating climate change. These transactions and commitments are largely conducted without the informed consent of local communities. Thus, large-scale land acquisitions (LSLAs), whether they are for conservation or commercial purposes, represent historical and concrete instances of perpetual and endemic dispossession, which result in a sense of collective insecurity and resistance. The section below outlines some key themes communities share in their struggle against KBPPA.

The Making of Dispossession and Resistance in the Southeast of Liberia

Senkwen and Juarzon are two communities in Liberia that are resisting the government's efforts to demarcate and regulate the KBPPA. The people of Senkwen and Juarzon have expressed concerns that the government's plans for the KBPPA will negatively impact their livelihoods and traditional ways of life. The conservation project is one of the largest in Liberia, spanning an impressive 290,167 hectares and making it the country's largest protected area. The protected zones are in the dense forest between the Grand Gedeh, Sinoe, and River Cess counties, representing more than 40 percent of the remaining upper Guinea forest corridors (Myers et al., 2000). To reduce global warming and protect biodiversity, Liberia has pledged to preserve 30 percent of its forest resources, leading to the creation of several national parks. The country has over ten conservation projects, covering over 10 percent of its land or 26.11 percent of the forest. (See Map 5.1 for a list of existing and planned conservation areas.)

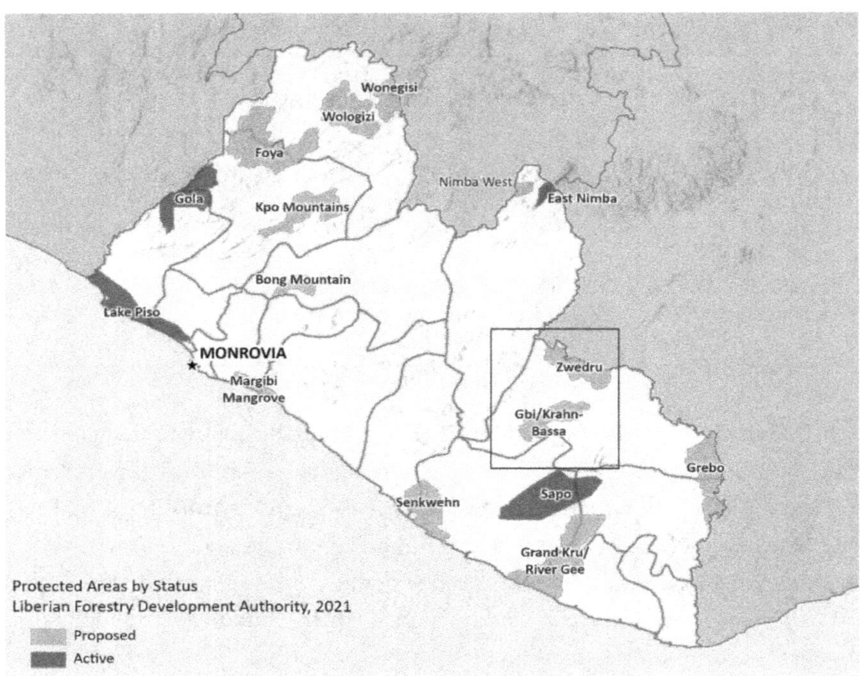

Map 5.1 Designated and proposed protected areas in Liberia. Data source: IUCN and UNEP-WCMC 2016 (Neugarten et al., 2017: 19).

Conservation is not new to Liberia. In 1953, the country passed its first conservation law, the Forest Conservation Act, which recognized forest resources as one of the most valuable natural resources in the country. The act required national parks, nature reserves, and wildlife management, regulating access to forest resources. In the 1960s, Liberia began categorizing its "national forests" with help from international partners. The Forestry Development Authority (FDA) was established in 1976 to promote and support activities protecting wild nature and natural resources.

The FDA manages Liberia's forest resources, including promulgating and enforcing forestry laws. Liberia—through the FDA—has implemented a number of forest sector reforms to promote sustainability, alleviate poverty, and improve livelihoods. The reform strategy is based on the "3C approach," which allocates the nation's forest resources most efficiently between commercial use, conservation, and community purposes (O'Hagan, et al., 2020). The Forestry Reform Law of 2006 and the Community Rights Law of 2009 aim to improve forest governance, strengthen community rights, and increase transparency in forest resource allocation. Furthermore, participatory forest resource management has been emphasized through conservation safeguards such as the National Forest Policy of Liberia and Community-Based Forest Resource Management (CBFRM).

Liberia is also involved in the carbon exchange scheme, which allows developing countries such as Liberia to generate carbon credits through conservation, thereby offsetting the emissions of industrialized nations.[7] These strategies aim to create jobs, drive economic growth, and generate revenues from the global market for carbon credits while also engaging in sustainable forest management, which scholars and activists refer to as the commodification of nature (Smessaert et al., 2020).

While the forest reforms were intended to benefit Liberians, they have not been without power struggles and elite capture. Many communities, including Senkwen and Juarzon, have expressed dissatisfaction with the reforms implemented in their areas. The two communities comprise over thirty towns inhabited by over 30,000 people from three different clans. Most community members depend on forest resources for their livelihoods and food security. In many towns, young people actively engage in small commercial activities, including farming (crops such as rubber, palm, coconut, and kola nuts) and artisanal diamond and gold mining. Also, artisanal mining and the charcoal industry—largely produced in rural areas—employs and supports between 100,000 and 1.5 million people (World Bank 2015, 2019). Furthermore, non-timber forest products such as wild fruits and nuts, honey, thatch, sticks, and medicinal roots/plants are vital for household and commercial purposes. These resources and local industries are critical to the local economy and provide food security and various income opportunities for community members.

Additionally, residents use the forest for cultural rites and recreation activities. In their view, demarcating the protected area will limit their access to these resources and effectively exclude them from their livelihoods. The major concern is the "one-sided" approach to these projects, including constricted opportunities to either consent or dissent, which leaves them powerless. Notably, most community members report that they have been dislocated from their lands or restricted from accessing resources. Land and resource loss, as well as the lack of consultation, has become a cause of frustration and resentment for many community members, particularly young men who rely on natural resources to make a living.

---

[7] Carbon trade can involve bilateral agreements, private parties, or market frameworks, like REDD+ under the United Nations Framework Convention on Climate Change (UNFCCC). Several scholars have criticized carbon trade, ranging from its danger of commoditizing nature to its practical implementation (Sovacool, 2011).

Senkwen and Juarzon Resistance against KBPPA

The Senkwen and Juarzon communities have been fighting for their land rights against government-sponsored conservation parks. Over the years, they have used different tactics to assert their rights. In 2018, the Senkwen people wrote letters to the FDA through their district leaders challenging the expansion of the Cestos-Senkwen portion of the KBPPA.

Their objection was that the park's expansion would encroach on their ancestral lands and deprive them of their source of livelihood. They complained that the government did not consult them before establishing the park. The letter also conveys that the creation of the park did not address their livelihood issues, which raises questions concerning the legality, human rights, and effectiveness of national and international conservation norms. "We don't know about the FDA owning this land or how it was turned into a [national] park. The forest rangers are blocking us from accessing our land, and they don't talk to us about our rights or livelihoods," says a local youth frustrated with the FDA. The claims emphasize the respect of customary land rights, which have recently gained national and international recognition and legal protection (the Land Rights Law, 2018), providing a basis for land rights advocacy.

Others complained that they have not been compensated for the loss of their traditional rights, including the loss of access to forest resources, land for farming, and artisanal livelihood. In Senkwen, a 50-year-old hunter protested:

> I used to get money from hunting, but now we are afraid to go into the bush because they will take our hunting materials (guns and bullets), including our dry meat. They can stop us anywhere in the bush (forest). The bush has no clear line of demarcation. I don't know where the lines are, so the rangers can take what we have and say we're breaking the law.

In frustration, a woman in Senkwen asked, "they told us not to go into the bush . . . How will we get food, medicine, and firewood?" An elder added: "In the past, we had access to our land and resources; animals were plentiful; we got wood, clean water, and thatch for our roofs, but now we can't get these things, and the government tells us not to step into the forest. How will we feed our families? This is why young people don't like the FDA's conservation work."

At the same time, several community members alleged that FDA rangers and some community members were part of an organized "cartel" that controlled the hunting market, supplying bush meat to nearby cities such as Buchanan and Monrovia. This claim was confirmed by a CSO staff working with the community:

"It's an FDA operation with some community members participating. If you have money, you can hire a local hunter for a fee and bribe your way into the forest. This is a common practice. Community members have reported the incident to the FDA, but nothing has happened. It's a major source of frustration in the community." Regardless of whether these claims are accurate, they do provide the community with a narrative that can be used to delegitimize the FDA rangers. Various forestry laws and regulations require the FDA to recognize, respect, and protect the rights of local communities (Community Rights Law, 2009), which is also tied to donor support.

Withdrawal from Agreed Rules

Members of the community claim that the FDA did not reply to their letters or respond to their complaints regarding dispossession, denial of access to livelihood, and illegal activities by forest agents. As a result, the community refused to work with the FDA on key steps like demarcating and harmonizing boundaries, rules around managing protected areas, and designing livelihood projects. These steps are part of the government and international sustainability claims. In situations where land rights are unclear, as in rural Liberia, it is difficult for the FDA to regulate or manage protected areas without clarity on whose land it is and who should be consulted. Thus, refusing boundary harmonization is a way to resist external control. A Juarzon youth leader said, "We will not participate in boundary harmonization unless they [FDA] acknowledge our customary land rights and support us in demarcating our land with our neighbors before the FDA can come in." A CSO member working in the community reports that most members share this attitude, including a traditional leader who said, "We just get told what the government wants. They come in big cars and tell us [conservation] is good for us, but every day, we have forest ranger problems. They stop us from entering the forest. Since we can't do anything about it, our young men are frustrated. They will challenge it."

Moreover, the FDA needs help managing these vast forests. While most of the focus is on external resources from donors, community participation is essential for safeguarding protected areas, which gives them local legitimacy and sustainability. For example, the Community Rights Law of 2009 is a legal framework that provides guidelines for protecting local communities' rights. One of the key requirements of this law is conducting a socioeconomic survey, documentation of resources and livelihoods, and mapping boundary areas. These processes, such as mapping and surveys, are conducted with the participation of local communities, providing them with a solid basis on which to negotiate or deny access to land and resources. Implementing national and international laws and safeguards for land expropriation protects rural communities from unfair expropriation and ensures that governments act responsibly and transparently.

Direct Resistance: Sabotage and Militant Tactics

Despite complaints from the community, the FDA and its partners continue to expand protected areas, which sometimes directly threaten the livelihoods of community members. According to a community leader, "Our cultural shrines are restricted," leaving the community members feeling "disrespected" and dehumanized. These actions make it clear to local communities that to maintain their lives, cultures, and livelihoods, they must resist conservation, whether by passive means or through active confrontation. In the Juarzon community, a group of young men organized a protest against the FDA rules regarding the forest. The group believed that the FDA's rules were limiting their ability to utilize the forest and that the rangers were not operating lawfully. The protest was peaceful, with the group gathering to make their voices heard. The young men hoped that their actions would bring about change and that the FDA would reconsider their rules and operations in the forest.

Further, in the Juarzon community, a group of young men accused the forest rangers of dealing with certain community members and outsiders to extract forest resources such as bushmeat and minerals. "We organized ourselves to protect our land and livelihood," a young man explained. This has resulted in conflict between different factions within the community and FDA rangers. Another group in the community also organized themselves to challenge this arrangement, ultimately leading to violent fighting and the death of some forest rangers. A local activist confirmed that the protests were unplanned, snowballing from isolated incidents into communal violence that quickly took off. This is not the first time protests have erupted in the community, "but the scale of the violence this time was unprecedented," he added. The local authorities were caught off guard by the sudden outbreak of violence, and it took several hours for them to bring the situation under control. In the end, dozens of people were arrested, and many more were injured. Therefore, community resistance to LSLAs does not need to be planned from above but can be triggered by collective frustration, making mobilization easier.

In River Cess, another area designed for conservation, a hunters' association publicly denounced the FDA in a community-wide meeting, accusing the agency of disrupting their livelihood. "We will not respect the rules not to hunt in the forest. They will have to bring the police or military," said a hunter. In a sense, this reflects frustration among youth, women, and elders who feel that state intrusion into their community undermines their authority, effectiveness, and relevance. This is different from the past when only chiefs and privileged elders made decisions, which created a cleavage between youth and elders. These revelations also indicate that when a community's interests are threatened, various stakeholders can collaborate to overcome the collective action challenge. Therefore, young people and elders in rural areas may adopt mutually reinforcing but flexible resistance strategies within the collective action framework. Collective action refers to group action whenever two or more individuals are involved.

Olson (1971) argued that individuals do not benefit from collective action unless a selective incentive directly benefits them. In other words, individuals are more likely to act in their self-interest unless there is a clear benefit to acting collectively. Yet, some confrontations need not be direct or collective. Occasionally, a threat from the community is sufficient to demonstrate resistance. According to an elder in Sankwen, "we [simply] inform FDA agents that we (elders) are not responsible for their safety, as the forest is large and dangerous. They understood our message. Some of the rangers have left our community, but they are still in the forest." Even where such a threat is not credible, the government is reminded of the Civil War, fought mainly by rural youth (Unruh, 2009). This history serves as a reminder of the potential for active opposition and threats to government initiatives.

Conclusion

This chapter examines Liberia's conservation efforts, focusing specifically on the KBPPA. The chapter explores the background of KBPPA, including the reasons for its establishment and the goals that it aims to achieve. It also looks at the processes involved in creating the KBPPA, including the role of different stakeholders, contested interests, and the challenges encountered. The findings suggest that the government's conservation efforts have been top-down, like LSLAs for agricultural concessions, and aim at protecting the environment and preserving natural resources. This type of conservation is inherently problematic; it is neither sustainable nor ethical.

Indeed, if conservation projects do not address the diverse needs of local communities, especially young people who depend on land and resources for survival, conflicts may arise between communities and conservation organizations. Some communities, such as the Senkwen and Juarzon, are deeply engaged in discursive resistance against KBPPA, using passive and active actions and strategies to challenge conservation practices. Furthermore, when young people do not have access to resources and cannot support their livelihoods, it may lead to poverty and economic hardship. Through our work, we have seen firsthand how this process can have far-reaching consequences for communities, individuals, and groups. Therefore, conservation programs should carefully consider local communities' land rights, respecting and incorporating those rights into decision-making processes. Also, it is necessary to analyze the various types of claims and interests within the community, which requires a comprehensive framework that includes discursive analysis and mapping the many factors affecting the community's economic and social well-being. In other words, if we ignore the diverse needs and aspirations of the community, protected areas or conservation projects become an imposition on the poor to cover the excesses of the rich.

References

Adams, W. M. 2004. *Against Extinction. The Story of Conservation*. London: Earthscan.

Agrawal, A. 2003. "Sustainable Governance of Common-Pool Resources: Context, Methods, and Politics." *Annual Review of Anthropology* 32: 243–62.

Alden-Wily, L. 2007. "So Who Owns the Forest?—An Investigation into Forest Ownership and Customary Land Rights in Liberia."

Boone, C. 2014. *Property and Political Order in Africa: Land Rights and the Structure of Politics*. London and New York: Cambridge.

Borras, S., and J. C. Franco. 2012. "Global Land Grabbing and Trajectories of Agrarian Change: A Preliminary Analysis." *Journal of Agrarian Change* 12: 34–59.

Brockington, D., R. Duffy, and J. Igoe. 2008. *Nature Unbound. Conservation, Capitalism and the Future of Protected Areas*. London: Earthscan.

Community Rights Law. 2009. An Act to Establish the Community Rights Law of 2009 with Respect to Forest Lands. Republic of Liberia.

Dressler, W., B. Büscher, M. Schoon, D. A. N. Brockington, T. Hayes, C. A. Kull, J. McCarthy, and K. Shrestha. 2010. "From Hope to Crisis and Back Again? A Critical History of the Global CBNRM Narrative." *Environmental Conservation* 37(1): 5–15.

Dudley, N., N. Ali, and K. MacKinnon. 2017. "Natural Solutions: Protected Areas Helping to Meet the Sustainable Development Goals." Briefing, Gland, Switzerland, IUCN World Commission on Protected Areas

Gilfoy, K. 2015. "Land grabbing and NGO advocacy in Liberia: A deconstruction of the 'homogeneous community'." *African Affairs* 114: 185–205.

Hollander, J. A., and R. L. Einwohner 2004. "Conceptualizing Resistance." *Sociological Forum* 19: 533–54.

Holmes, G. 2007. "Protection, Politics, and Protest: Understanding Resistance to Conservation." *Conservation and Society* 5(2): 184–201.

IUCN. 2019. International Union for Conservation of Nature Annual Report 2019. https://portals.iucn.org/library/node/49096

Land Rights Law. 2018. An Act to Establish the Land Rights Law of 2018. Republic of Liberia.

Liberian Extractive Industries Transparency Initiative (LEITI). 2014. Scoping Study on the Mining Sector (Liberia).

Moyo, S., P. Jha, and P. Yeros. 2019. "The Scramble for Land and Natural Resources in Africa." In S. Moyo, P. Jha, and P. Yeros (eds.), *Reclaiming Africa: Advances in African Economic, Social and Political Development*, 3–30. Springer.

Myers, N., R. A. Mittermeier, Cristina G. Mittermeier, Gustavo A. B. da Fonseca, and Jennifer Kent. 2000. "Biodiversity Hotspots for Conservation Priorities." *Nature* 403(6772): 853–8.

Neugarten, R., M. Alam, N. Acero, M. Honzák, D. Juhn, K. Koenig, T. Larsen, K. Moull, A. Rodriguez, T. Wright, L. Walsh, J. Donovan, P. Mulbah, J. Valenza, K. Reuter, R. Portela, T.

Noviello, S. Wade, D. Hole, and J. Junker. 2017. Natural Capital Mapping and Accounting in Liberia: Understanding the Contribution of Biodiversity and Ecosystem Services to Liberia's Sustainable Development.

O'Hagan, R., A. O'Connor, J. Fa, and T. Sunderland. 2020. *Community Forestry in Liberia: Progress and Pitfalls*. Cambridge: Cambridge University Press.

Oldekop, J. A., G. Holmes, W. E. Harris, and K. L. Evans. 2016. "A Global Assessment of the Social and Conservation Outcomes of Protected Areas." *Conservation Biology* 30(1): 133–41.

Olson, M. 1971. *The Logic of Collective Action: Public Goods and the Theory of Groups* (Second printing with a new preface and appendix). Harvard University Press.

O'Mahoney, J. 2019, March 22. "Liberia's New Land Rights Law Hailed as Victory, but Critics Say It's Not Enough." *Mongabay*. https://news.mongabay.com/2019/03/liberias-new-land-rights-law-hailed-as-victory-but-critics-say-its-not-enough/

Republic of Liberia. 2006. National Forestry Reform Law of 2006. www.fao.org/forestry.

Sawyer, A. 1992. *The Emergence of Autocracy in Liberia: Tragedy and Challenge*. San Francisco: Institute for Contemporary Studies.

Scott, J. C. 1985. *Weapons of the Weak. Everyday Forms of Peasant Resistance*. New Haven and London: Yale University Press.

Smessaert, Jacob, Antoine Missemer, and Harold Leverel. 2020. "The Commodification of Nature, a Review in Social Sciences." *Ecological Economics* 172: 106624.

Sovacool, B. K. 2011. "Four Problems with Global Carbon Markets: A Critical Review." *Energy & Environment* 22(6): 681–94.

Stevens, C. 2014. "The Legal History of Public Land in Liberia." *Journal of African Law* 58(2): 250–65. https://doi.org/10.1017/S0021855314000059

UNESCO. 2008. Links between Biological and Cultural Diversity-Concepts, Methods and Experiences, Report of an International Workshop, UNESCO, Paris.

Unruh, J. D. 2009. "Land Rights in Postwar Liberia: The Volatile Part of the Peace Process." *Land Use Policy* 26(2): 425–33.

World Bank. 2015. A National Biodiversity Offset Scheme: A Road Map for Liberia's Mining Sector.

World Bank. 2019. Libera Forest Sector Project: Opportunities for Charcoal and Sustainable Forest Management.

The Boeung Kak Lake Evictions and the Experience of Women in Cambodian Land Dispossessions

Mariko Frame

Introduction

Cambodia, a poor nation whose GDP per capita is 1785.6 in current US dollars (World Bank, 2023), is arguably a peripheral country in the capitalist world-system. As discussed in the introduction of this book, peripheral countries in the capitalist world-system are structurally plagued with ecological imperialism in the form of extractivism, socio-ecological degradation, and outflow of resources and profits to foreign investors, often with the collusion of domestic elites. Like elsewhere throughout the Global South, land grabs in Cambodia are inextricably linked to a neoliberal conception of development that heavily favors foreign investment and export-led development as engines of economic growth. In the early 2000s, with the financial and technical support of the International Financial Corporation (the private sector arm of the World Bank), the Cambodian government amended its 1994 Investment Law to encourage foreign direct investment (FDI) inflows. This triggered sharp increases in FDI, including agriculture and natural resource extraction, with major repercussions for land concessions (Oakland Institute, 2013). Within the Southeast Asian context, Cambodia appears to be subject to a "regional" ecological imperialism, as investors in land concessions are largely driven by Southeast Asian (Vietnamese, Singaporean, Thai, Malaysian) and East Asian (Chinese, Korean) capital, as these countries pursue their own paths of industrialization and increased resource extractivism both domestically and overseas (Frame, 2018).

Over the past two decades, the government of Cambodia has forcibly removed hundreds of thousands of Cambodians from their land, resulting in civil unrest, human rights abuses, deforestation, and socioeconomic crises. As elsewhere throughout the Global South, key actors include the state and military, investors both foreign and domestic, and also the influence of international financial institutions. Officially, economic land concessions (ELCs) are justified as needed for development. In reality, the land is often handed over to foreign commercial enterprises that pay off prominent government figures or their relatives. The vast majority of ELCs have been issued in violation of Cambodia's 2001 Land Law and its sub-decree on ELCs. Their requirements regarding size, prior environmental and social impact reports, prior consultations and consent of affected communities, transparency, and fair and adequate compensation have been routinely ignored. Efforts to enforce the requirements in civil lawsuits have been met with years of court inaction or retaliatory criminal charges. Although numerous local and international NGOs, foreign governments, and international organizations have criticized the practice of land grants, and the government issued a moratorium in 2012 on ELCs, the evictions have continued.[1]

This chapter presents a case study compiled from interviews of twenty-three women and one man (who was present with his interviewed wife) affected by the Boeung Kak Lake evictions, capturing the experiences of women evictees in terms of resistance and political repression, economic consequences, as well as the toll such evictions have had on their psychological and physical health. Despite being a small case study, it nonetheless reiterates the findings of other studies that highlight the gendered toll of land grabbing. As elsewhere throughout the Global South, women in Cambodia are actively involved in protesting and resisting land grabbing through activities such as riots, roadblocks, trespassing, and damaging company property (Park, 2019). Likewise, women have been at the forefront of the Boeung Kak Lake movement, leading marches and delivering petitions to government ministries despite regular arrests and violence from security forces. Renowned activist leaders, such as Tep Vanny, have campaigned on behalf of others and demanded justice for those that have been arrested and also traveled internationally to raise awareness (LICADHO, 2018). Tep Vanny and the Boeung Kak 13—a prominent group of women activists who resided around Boeung Kak Lake—employed a wide variety of resistance tactics, such as refusing to move from their homes, protesting on the streets of Phnom Penh, using national legislation, submitting complaints to the World Bank, and

---

[1] Please see the website of LICADHO, which has a host of reports documenting such trends at https://www.licadho-cambodia.org/topic/land

utilizing humor such as donning bird nests on their heads to defend their roles as "mother hens" (Baaz ., 2017).

However, as stated in the UN CEDAW[2] Committee's 2019 Concluding Observations on Cambodia's official report, women human rights defenders, trade union workers, and land and environmental activists continue to face harassment and intimidation. LICADHO, one of Cambodia's main human rights organizations, has documented how women land campaigners and human rights defenders at the forefront of peaceful protests face state violence, imprisonment, judicial harassment, and arbitrary detention in Cambodia (UN Office of the High Commissioner for Human Rights, 2019).[3]

This small case study also underscores the worsening economic consequences for women that land dispossession frequently results in. The UN CEDAW 2019 report points out how women in Cambodia lack substantive equality with men, particularly in relation to ownership of land and access to adequate housing and access to economic opportunities in rural areas. The report further notes that the granting of the majority of land titles is to men and recommends that Cambodia allocate and distribute land so that women have equal access. It is therefore not surprising that economic outcomes for women are severe where land grabbing has been documented, including high levels of livelihood and asset harm (McGinn, 2015). Finally, this study documents the psychological, social/familial, and physical stress for the women interviewees due to the evictions. Overall, as elsewhere in the Global South, women are active and crucial actors in movements of resistance, yet continue to pay a heavy toll from the intersectionality between patriarchy, class, and the dynamics of capitalist accumulation.

Boeung Kak Lake Land-Grant Evictions

While the vast majority of land grabbing in Cambodia is for agricultural purposes, the Boeung Kak Lake evictions present an urban tale of land dispossession for commercial "development," illustrating how land grabbing can reconfigure not only rural but urban spaces along lines of unequal power, class, and gender dynamics. In the past, Boeung Kak Lake was the largest natural lake in Cambodia's capital Phnom Penh. An estimated 4,250 families lived on the lakeside, and farmers earned a living through cultivating water vegetables.

---

[2] Cambodia has been a state party to the UN Convention on the Elimination of all forms of Discrimination Against Women (CEDAW) since 1992.

[3] https://www.licadho-cambodia.org/articles/20181126/153/index.html

Water played a major role in the city's drainage system, as the vegetation naturally filtered water from the city's sewage. The lake supported a thriving tourist industry, as many guesthouses and businesses built on the shore of the lake catered to budget travelers. In 2007, Phnom Penh's government approved the subleasing of Boeung Kak and all the surrounding land to Shukaku Inc., an exclusive residential and commercial development project co-owned by Cambodian Senator Lau Meng Khin, with the shadowy involvement of foreign investors, including Chinese capital (Ruo, 2011). Shukaku reportedly paid a mere $.60 per square meter at a time when a standard apartment rented for $2,913 per square meter (Hayman and Rith, 2007). Although laws prohibit a foreign entity from owning land or substantially altering it, the company permanently evicted all residents and other businesses, then proceeded to destroy the lake by filling it with sand.

In 2015, I conducted a set of interviews in collaboration with NGO-CEDAW and LICADHO, two prominent civil society organizations in Cambodia. The intent of the interviews was to publish a follow-up report on the economic, political, and social conditions of women Boeung Kak Lake evictees both prior to and after their evictions. While the report was not ultimately published due to the political sensitivity of the subject matter, I received approval to publish the results as academic work.

Interview Process

With the help of a Khmer translator provided by the organizations, we interviewed twenty-three women (and one man, a husband who was present with his wife during the focus group interview) involved in evictions from Boeung Kak Lake. Six of the interviewees were not, in the end, evicted from their homes and land, though one of them did lose part of her land. Seven of the interviewees were evicted from their homes and land and took the monetary compensation offered to them by the company. The upper limit on the monetary compensation was $8000 plus an extra 2 million riel (approximately 485 US$), but not all of the seven interviewees received this full amount. This study also undertook a focus group of interviewees who had taken the relocation site option offered to them by the company. Aside from monetary compensation, evictees were also offered the option to have new homes provided for them at a relocation site some twenty-three kilometers from Phnom Penh. To interview the focus group, we traveled to the relocation site, known as "Seven N G" and held an interview with seven people in the home of one of the interviewees. All interviews began with an open question asking the interviewees to explain their personal stories of how the evictions took place or almost took place. The interviews then went into more specific questions, particularly concerning the overall economic impact of the evictions on the interviewees. Questions were asked concerning wealth lost, livelihoods and incomes lost, and changes in expenses. While the major focus of this study was on the economic impact of the evictions, questions on the political, social, and emotional impact were also asked. (*Names were changed to protect the identity of the interviewees.*)

State/Private Security Forces, Resistance to Evictions, and Political Repercussions

All interviewed evictees were active in protesting and resisting their evictions while facing severe repercussions. All of the interviewed evictees reported that their eviction had been forced or at least forcibly attempted, underscoring how the threat of the state, and in the case of Boeung Kak Lake, the threat of the state in collusion with corporate private security forces, is a central component in land dispossession. The majority of evictees reported that the company gave them only a seven-day notice and threatened that if they did not move their belongings and dismantle their houses within seven days, their houses would be destroyed and neither the company nor the government would be held responsible. As one interviewee stated,

> At first, the company pushed people to take the ($8000, 2 million riel) offer, and respect state law. People didn't want to take the offer, but security forces came in the night, carrying guns. One night I smelled gasoline, and I got scared. I saw other people's houses being destroyed. I was forced to take the offer, and in the end, did not even get the 2 million riel, only the $8000. —Chantrea

Chantrea went on to describe how, as a result of the evictions, she was forced to work as a dancer in an entertainment club—a troubling statement given Cambodia's track record with sexual and gender-based violence, and lack of law enforcement protection in regards to such issues (LICADHO, 2021).

> The money was not enough to buy land. My grandmother got sick, so I had to use the money for her, and the children. I now have to work as a dancer at an entertainment club. —Chantrea

Another interviewee said,

> The company gave us notice to leave within seven days. They told us if we don't leave, then our belongings are in danger and they will not be responsible. I went with others to the Phnom Penh Municipal Hall to protest, and tried to talk with the officers there. I've been arrested and in jail for one month and three days, and beaten, on the legs. The company security guards beat those who protested on company land. —Bopha

The majority of the women interviewed—evicted, non-evicted, and the focus group—protested multiple times since the beginning of the evictions. However, the cost of such resistance in Cambodia can be severe with ongoing sociopolitical repercussions. Five of the adult interviewees and one child reported being beaten and in some cases quite severely. A couple of the interviewees reported being arrested or of knowing close friends or activist representatives who were arrested. Seven of the interviewees reported harassment and intimidation by company security, government security, and local officials. Harassment included feeling discriminated against by local authorities and feelings of being followed and harassed when trying to sell their goods as vendors. Many of the interviewees stated that government authorities framed their protests as being instigated by the opposition party. In sum, all interviewees report that the company forced, or attempted to force, eviction through intimidation by company security, underscoring the collusion between state and private security forces.

The Socioeconomic Impact of Eviction

> Before the eviction I had my own land and house. My land was 6m x 100m and my house had ten rooms. I was able to get some income from renting rooms, and I owned a laundry shop. From rental and laundry I could earn between 20,000–40,000 riel a day. I was notified that I had seven days to take the $8000, 2 million riel compensation, or the house would be destroyed. When I saw many security guards from the government and company, I was shocked, and fainted. I took the money, and gave it to my husband to take care of our children. He left me for another woman. I have no home now, and sleep on the street. When it rains, I sleep under the roofs of buildings. I now collect rubbish on the streets, and sometimes do laundry. The income is not regular, and if I can get hired, I can maybe get $2.50 in a day. —Chanlina

As elsewhere throughout the Global South, foreign investment and foreign investment in collusion with domestic capital, in sectors such as agriculture, extractivist sectors, and land, have driven environmental degradation and destruction of livelihoods not captured by the corresponding growth in GDP or official stories of economic growth and development. Cambodian women in particular face challenging economic circumstances. As of 2018, only about 15 percent of adult women completed secondary education. Women are overrepresented and vulnerable in the informal sector, including being employed in low- and unstable-wage jobs, risky, and casual work environments. Poor and rural women tend to be saddled with higher rates of domestic work, home-based work, and as smallholder farmers. While women owned 65 percent of all businesses in Cambodia, those businesses are on average smaller and less profitable than businesses run by men (Oxfam in Cambodia, 2020). In the case of Boeung Kak Lake evictions, economic outcomes for all women interviewees, both in terms of livelihoods and wealth, worsened considerably both quantitatively and qualitatively.

Most of the evictees were offered the choice of either accepting $8,000 and 2 million riel as compensation for their lost houses and land—no matter the size of the land or the house. Or, they were given the opportunity to be relocated and given a new house, with 2 million riel as moving compensation. Two of the adult interviewees (one of which was a family with one child), however, were not offered the $8,000, 2 million riel compensation. One very poor interviewee owned a small piece of land (5 m × 8 m) and a cottage, with a roof and no walls. The interviewee claims not to have been given a notice but was only told by neighbors. When she inquired, she found out that her name was not on the list to receive compensation. When she confronted the company, the company refused to pay compensation, and only after many confrontations, they agreed to pay a mere $100. The interviewee refused this small amount and in the end, got nothing. The other interviewee who did not receive the full $8,000 compensation owned a house and land (5 m × 6 m) before being evicted. She was given a notice to leave within seven days, but she did not leave. According to her, the company pumped sand and destroyed her house. When she confronted the company, the company argued that her house did not have a strong foundation and used this as an excuse to only offer $500 as compensation. She refused and in the end, got nothing.

Livelihoods and Income Lost

With the exception of one or two of the interviewees, all of the women interviewed had worked on or near the Boeung Kak Lake area, earning their livelihoods through a variety of mostly small, informal businesses such as street vendors, sewing, or renting rooms from their houses. As such, almost all of the interviewees were dependent on the flow of customers and business in the Boeung Kak Lake area to support their livelihoods. Almost all of the interviewees, who could be described as low to low-middle income, felt that before facing eviction, they had made enough money to at least survive and pay for basic daily necessities. This situation changed drastically with eviction. As the focus group told us,

> Before the eviction, we weren't rich, but we had enough to live off of. We had small jobs, like selling vegetables on the streets as vendors, but we had enough to send our children to school. Now, at the relocation site, we can't bring in enough money. There are too many sellers here, and not enough buyers. Everything is much more expensive than it is in Phnom Penh. And our families are forced to live separately from each other and it is expensive traveling to and from Phnom Penh. It is very difficult. We earn less, but face higher expenses. It is really hand-to-mouth living. Before, some of us used to grow food on our land by the lake, some of us used to raise pigs, fish, or raise lilies to sell. Without enough income, we had to borrow money from the banks. But without enough income, we were not able to pay back the banks, so the banks confiscated the new homes given to us at the relocation site. They bought out new homes for lower prices than they sold them for. Now we are renters, and must continue to borrow money, and pay interest. Many of us had to pull our children from their education, and they have to work in factories nearby. One of us has a smart, pretty daughter who is sixteen. She used to go to school, but now she works in a dance club, and we worry about her future.

With the exception of one of the interviewees, all reported either complete loss of livelihoods or such degradation of incomes that they felt desperate and constantly worried about finances. As discussed, many of the interviewees worked in small businesses dependent on the Boeung Kak Lake area. Ten of the interviewees, plus those in the focus group, said that the development of the lake area had resulted in fewer customers. Their businesses were no longer sustainable. The focus group in particular complained that business at the relocation site was unsustainable; being so far from Phnom Penh, in an area with a limited population, resulted in too many sellers and too few buyers. The loss of income was devastating for the interviewees. A number of the interviewees from the focus group reported having to withdraw their children from schooling so that they could work in factories. The loss of wealth, in land and housing, of the interviewees was also directly linked to the deterioration of income for some of the interviewees. Three of the interviewees, plus members of the focus group, reported losing income as they used to rent out rooms in the houses they lost or used to farm or fish and sell their products as small vendors. Other interviewees reported a loss of livelihood and income as a repercussion of their political protests. One interviewee, who frequented a number of protests and did not, in the end, lose her land, nonetheless lost her means of income. She reported that she could no longer sell as a vendor on the streets as she used to, as since the protests she is harassed by the authorities. Another interviewee reported that she was repeatedly shocked with an electric baton during the protests. She felt her health had suffered greatly from these electric shocks and said she was no longer able to work.

Wealth Lost

Land eviction often results in the considerable loss of wealth for evictees. Being low- to lower-middle-income group, loss in wealth can have devastating repercussions throughout their whole lives. Obviously, the land itself is appropriated, but with the land, housing, gardens, and other material goods are lost. While evictees sometimes receive some form of compensation, either in monetary form or through the allocation of new land and houses, a common grievance is that such compensation does not actually equal the amount of wealth lost.

Nineteen interviewees suffered eviction from their land and loss of houses. All held serious grievances concerning the extent of the wealth they lost. The wealth owned before eviction varied considerably among interviewees, with some owning just small plots of land, to others owning over an acre. Houses owned varied considerably as well, with the poorest of the interviewees having owned only a simple shack without walls to one of the wealthier interviewees that owned a house with ten rooms that brought in considerable rent as her income.

As for the relocation site focus group, all lost land and homes and were only compensated $500 dollars as remuneration for moving expenses. While the focus group interviewees did receive new houses at the relocation site, the interviewees complained of their houses being badly constructed, made of cheap material, and prone to leaky roofs. More disturbingly, many of the nine focus group interviewees eventually lost their new homes to banks. The focus group interviewees all complained of insufficient incomes at the relocation site. Unable to sustain their livelihoods, many of the interviewees borrowed money from banks. In time, unable to repay the banks with their insufficient income, the banks then confiscated their new homes (at prices lower than market value), and now the interviewees have once again lost their homes. As a result, the interviewees have found themselves without house ownership, having to pay rent while living on insufficient income and continuously in debt. Overall, all interviewees suffered considerable losses in land and housing with arguably inadequate compensation.

Many of the (evicted) interviewees also reported higher expenses. Five of the interviewees, plus the interviewees of the focus group, reported struggling to pay rent. Before, they used to own their houses and land, but with insufficient compensation to pay for new homes, they are forced to rent rooms. The other (evicted) interviewees reported having to live with family, or, in the sad case of one elderly woman having to live on the streets. Two of the (evicted) interviewees plus ten interviewees from the focus group reported being in debt as a result of financial hardships related to the evictions.

Six of the twenty-four interviewees were not evicted, though one did lose part of her land. Though these six interviewees were not evicted, their homes and land were initially included in the project development area and under threat of eviction. All six were heavily involved in protests, and through the process of protesting and pressuring the government and company, they claimed to have forestalled eviction. However, it is likely that this was possible as many of them claimed their land was not directly on the lake but further in land. While these women suffered considerable stress, and some of them were imprisoned and beaten, they did not, in the end, lose their land or homes. However, one of these five interviewees complains that her house now suffers from flooding.

The Social, Emotional, and Health Impacts of Eviction

Living at the relocation site, we are so far from our families now. Our families are dispersed. We are separated from them, and it is very hard on us. Some of our husbands have been really hurt by the eviction. They get angry, and have become drunkards. There is violence in our homes. We feel hopeless and depressed. — Focus Group

My children are always afraid for me (since I protest). My neighbors are on the side of the village chief—they accuse me of trying to get money. I feel very sad for those (activists) in jail. I feel hopeless, and have lost trust in the government. The government treats us as enemies, as part of the opposition party. —Champei

The social impact of eviction, and more specifically, of protesting, varied among the interviewees. As stated above, most of the interviewees were involved in protests. While a few of the interviewees reported having felt supported by their families and neighbors, a number felt discriminated against by their local communities. Seven of the interviewees reported being mocked by their neighbors or accused of being part of the opposition party. Five of the interviewees reiterated feeling discriminated against by the local authorities. Three of the interviewees reported having no support for their families. And five of the interviewees reported a worsening of family relations resulting from separation from their families or increased arguments in the household. The focus group also reported issues of familial separation and feeling discriminated against by local authorities.

The eviction had a harsh toll on the emotional well-being of the interviewees. The interviewees who did not, in the end, become evicted still reported living in fear of losing their houses and feelings of distrust toward the government. Though some of the non-evicted interviewees reported feelings of defiance and empowerment through their protests, they said they felt great concern and sorrow over their fellow activists who were still in prison. The emotional impact on the evicted interviewees, both those at the relocation site and those who took the compensation, was much worse. Almost all reported feelings of great pain and hurt, fear of losing their lives after conflict with authorities, hopelessness, stress over finances, mental problems, and some even had thoughts of suicide, and some interviewees reported increased health concerns and domestic violence. The evictions caused many of the interviewees to endure a decreased standard of living, including much more pollution and much poorer sanitation. The relocation sites often had no running water, no sanitation, no electricity, no medical facilities, and little vegetation to keep the dust at bay. This fostered an environment that contributed to respiratory illnesses and other transmittable diseases.

Such socioeconomic outcomes demonstrate a major weakness in the logic often found concerning large-scale land concessions found elsewhere throughout the Global South in land laws—that such concessions can be adequately compensated for and that they should be considered legal if they are. In reality, a one-off payment cannot account for livelihoods permanently lost and the ensuing emotional, physical, and political distress. As has been found with the rural dispossession cases in this book, the value of land exceeds a monetary number, providing self-sufficiency and ecological connection while also being inextricably linked with social and familial relations that provide psychological and economic security.

Conclusion

While the official story of post-conflict Cambodia has been of economic blossoming and growth, driven by market principles such as export-led development and foreign investment, the socio-ecological realities, as elsewhere throughout the Global South and as documented in this edited volume, are far more distressing. In contrast to the sanitized models of growth found in orthodox economic models, economic "growth" in capitalist economies is undergirded by processes of capital accumulation that involve the forcible disciplining of labor and transfer of land and resources from one socioeconomic group to another. In other words, far from the neutral working of the market portrayed in orthodox models, in reality market processes rely on enforcement by extra-economic actors, of which the state is a key player. In Cambodia, the economic, social, political, and environmental consequences associated with ELCs are some of the most extreme in the world, with hundreds of thousands of Cambodians suffering from land dispossession. Historically, capitalism has always involved the dispossession of land and resources from one group to another, both at the global level through colonialism, and also within nations between classes. In Cambodia, state, private security forces, neoliberal policies emphasizing market-based growth through deregulation, foreign investment, and export-led policies have set the stage for the grabbing of land as a dwindling and precious commodity in an era of ecological crises. For women, they experience this ecological imperialism at the intersections of gender, class, and capitalism, pushing back as critical players in resistance, yet suffering gendered outcomes of political repression, loss of wealth and livelihoods, worsening economic outcomes, and social, psychological, and health stresses.

References

Baaz, M., M. Lilja, and S. Vinthagen. 2017. "Resistance Studies as an Academic Pursuit." *Journal of Resistance Studies* 3(1): 10–28.

Frame, Mariko. 2018. "The Role of the Semi-Periphery in Ecologically Unequal Exchange: A Case Study of Land Investments in Cambodia." In R. Scott Frey, Paul K. Gellert, and Harry F. Dahms (eds.), *Ecologically Unequal Exchange: Environmental Injustice in Comparative and Historical Perspective*, 75–106. London: Palgrave MacMillan.

Hayman, A., and S. Rith. 2007, February 9. *Boeung Kak Lake Latest City Sell-off*. Phnom Penh Post. Retrieved July 6, 2023. https://www.phnompenhpost.com/national/boeung-kak-lake-latest-city-sell

LICADHO. 2018, November 26. *Article: 16 Days of Activism Against Gender Based Violence: Tep Vanny*. Licadho. Retrieved July 4, 2023. https://www.licadho-cambodia.org/articles/20181126/153/index.html

LICADHO. 2021, May 12. *Statement: Take Legal Action on Cases of Domestic Violence, Sexual Harassment and Sexual Violence Against Women*. Licadho. Retrieved July 4, 2023. https://www.licadho-cambodia.org/pressrelease.php?perm=476

McGinn, Colleen. 2015. "'These Days We Have to Be Poor People': Women's Narratives of the Economic Aftermath of Forces Evictions in Phnom Penh, Cambodia." Paper presented at the International Academic Conference on Land Grabbing, Conflict and Agrarian–Environmental Transformations: Perspectives from East and Southeast Asia, Chang Mai University, June 5–6.

Oakland Institute. 2013. "World Bank's Bad Business in Cambodia."

Oxfam in Cambodia. 2020, February 25. *Promoting Economic Empowerment for Women in Marginalized Conditions | Oxfam in Cambodia*. Oxfam in Cambodia. Retrieved July 4, 2023. https://cambodia.oxfam.org/latest/publications/promoting-economic-empowerment-women-marginalized-conditions

Park, Clara Mi Young. 2019. "'Our Lands are Our Lives': Gendered Experiences of Resistance to Land Grabbing in Rural Cambodia." *Feminist Economics*, 25(4): 21–44. https://doi.org/10.1080/13545701.2018.1503417

Ruo, L. 2011, November 16. *A Shadow over Boeung Kak Lake (1)*. China Dialogue. Retrieved July 6, 2023, from https://chinadialogue.net/en/business/4629-a-shadow-over-boeung-kak-lake-1/

UN Office of the High Commissioner for Human Rights. 2019, November 12. *Committee on the Elimination of Discrimination against Women: Concluding Observations on the Sixth Periodic Report of Cambodia*. OHCHR. Retrieved July 4, 2023. https://www.ohchr.org/en/documents/concluding-observations/committee-elimination-discrimination-against-women-concluding

World Bank. 2023, April 12. *Overview: Development News, Research, Data*. World Bank. Retrieved July 4, 2023. https://www.worldbank.org/en/country/cambodia/overview#1

7

Community-Led Activism in Cambodia's Prey Lang Forest

Narith Nou, Mariko Frame, Felix Mantz, and Sam Grant

Introduction

Large-scale land investments in Cambodia have resulted in some of the most extreme examples of land grabbing in the Global South. According to the International Federation for Human Rights (FIDH), by 2014 some four million hectares of land had been invested in, with land grabs affecting at least 770,000 people (Flynn et al., 2021). For decades, Cambodia's national human rights NGO, the Cambodian League for the Promotion and Defense of Human Rights (LICADHO), has collected extensive documentation of the civil unrest, human rights abuses, and corruption underlying this trend.[1] Like elsewhere in the Global South, several key actors are typically involved in land grabs: the state and the military, along with both foreign and domestic investors. International financial organizations, in particular the World Bank, are also implicated. As elsewhere in the Global South, the World Bank—specifically its private sector arm, the International Financial Corporation (IFC)—provided the financial and technical support to create an investment climate that strongly favored foreign direct investment inflows, triggering a sharp increase in agriculture and natural resource extraction (Oakland Institute, 2014). And, like elsewhere in the Global South, such changes are justified under the auspices of "development." Land grabbing is driven overwhelmingly by foreign investments from East Asian and Southeast Asian nations, but investments from Cambodian companies also play a central role. The land grabbing phenomenon in Cambodia has been

---

[1] For economic land concession details and numerous briefings and reports on land grabbing, please see LICADHO's website, https://www.licadho-cambodia.org/

an ecological disaster; aerial footage collected by LICADHO shows a high correlation between land concessions and deforestation.

Prey Lang is one of Southeast Asia's last major lowland rainforests and home to some 250,000 inhabitants, many of whom are an Indigenous group known as the Kuy. Located in northern Cambodia, the forest covers approximately 5,000 square kilometers. Prey Lang, which means "our forest" in the Kuy language, is an important part of Kuy cultural and spiritual life. The Kuy people live primarily on the land, farming rice and crops, and rely on the forest as a source of income. Prey Lang is a biodiversity hotspot and includes seven different types of rainforest home to a number of rare and endangered species. Officially, 4,320 square kilometers of Prey Lang are protected as a wildlife sanctuary. However, deforestation continues to be driven by illegal logging, industrial agriculture, and mining, even in protected areas, and has intensified in recent decades (PLCN, n.d.a)

This chapter is the product of a dialogue between the editors and Narith Nou, a former project coordinator with the Prey Lang Community Network (PLCN). PLCN is a network of local community members working to save the Prey Lang forest. The network patrols the forest, records illegal logging and conversion of forest with the protected Prey Lang Wildlife Sanctuary, and communicates its findings to policymakers and the general public through reports and social media. PLCN is an example of grassroots, community-driven environmental resistance, with 400 active volunteer members from communities in Prey Lang's surrounding provinces. PLCN is the recipient of numerous international prizes, including the International Society for Tropical Foresters Innovation Prize, awarded by the Yale School of Forest and Environmental Studies, the United Nations Development Program (UNDP) Equator Initiative Prize, and others (PCLN, n.d.b).

Editors: Please introduce yourself. Please describe the kind of work you were doing with PLCN, and your position.

Prey Lang Community Activist: I've been with PLCN for seven years. Since 2014, I've been a project coordinator with PLCN until June 2021. In my role, I provided all types of support, including reporting, releasing statements, organizing campaigns, international campaigns, conducting meetings, organizing coordination committees to discuss the activities of PLCN, conducting core group meetings, annual meetings, organizing meetings with organizations like LICADHO (National Cambodian Human Rights Organization), CCHR (the

Cambodian Center for Human Rights), OHCHR, (the UN Human Rights Office in Cambodia) . . . . many, many meetings . . . . I also provided training, educating others in forest law, and provided training in Prey Lang APP, an APP used to calculate data on the Prey Lang forest. I did everything for PLCN . . . .

PLCN is a strong network, a movement, and I always analyzed the politics for them and provided strategy for specific political contexts. I always made plans for them. PLCN helps find good ways to mobilize the volunteers for the Prey Lang forest for any activity, any campaign, to help protect the volunteers, so that they don't suffer from their activities.

Editors: What does PLCN stand for?

Prey Lang Community Activist: Prey Lang Community Network. We provide the people fighting for Prey Lang forest with advice about how to make their network stronger, how to make the network withstand pressure and challenges from the authority or the government, to make sure that the volunteers can keep up with their activism.

One example of this work is the Prey Lang Annual Ceremony in 2020. One week after PLCN sent a letter to MoE (Ministry of the Environment) to invite and inform them about the annual ceremony, the MoE announced organizing the Prey Lang ceremony by themselves. They did not care about PLCN's letter. This means they don't want PLCN to do any activities in Prey Lang's protected area controlled by MoE. So I and PLCN together analyzed what will happen at the PLCN's annual ceremony. They will block, crackdown, and arrest people if PLCN wants to achieve its plans. PLCN decided to organize the ceremony. When MoE's rangers come to block it, PLCN will negotiate. If this is not possible, they will go back to their village and release a statement about this case. PLCN explained this strategy to their members so that they know they should not face with the MoE at this time. When they know about pressures and challenges in advance, members can adapt to those things. If they don't, PLCN members will be scared, lose control, and maybe leave the network.

PLCN doesn't have one leader; it has one hundred leaders. If one leader leaves, there are many others. Legal analysis is very important, acting under Cambodian law. We need to know how everyone working with PLCN has rights under Cambodia law. PLCN has the best approach for its members. They know how to divide responsibilities and coach all core groups and active members. Whatever the leader can do, other core groups and active members can also do. Any activities have been invited by partners, it's not only for leaders. When

NGO partners invite PLCN to join any activities, the leader will discuss with other core group members who could join those activities and can come back to share them.

Editors: How did you get involved with PLCN? What made you want to do this kind of work?

Prey Lang Community Activist: This work is very hard for the people, also for me it was very hard. You can see my hair has changed, it has become white because of too much stress! Why did I work with PLCN? Because I love to work with the community, with the movement, with the grassroots people. I study psychology and law, so I need to know about the mind of the people and our rights.

Working with PLCN is a hard responsibility, but I love to improve my abilities, my capacities to protect the forest. I've always been a student-activist. Since 2008, I've been an environmental and student-rights activist.

Editors: What are the biggest environmental challenges facing Cambodia today? What are they primarily being driven by?

Prey Lang Community Activist: The biggest challenge facing Cambodia's environment to date is development—"development" according to the Cambodian government. Such development includes taking the land's protected areas, natural lakes, rivers, and seas and providing them to the private companies. The social land concessions[2] are a big challenge for Cambodia, not only for the forest but also for the people living around the forest, and the people living in the city, in the town. When a company comes, they want the land, they want the forest, so the government provides the license to the company. They don't care about the poor people, about the owners of those lands or those forests. They don't respect the rights of the people; they can do whatever they want.

Editors: Who (as in, what type of groups) are driving the environmental degradation?

Prey Lang Community Activist: In Cambodia, with the timber concessions, we have the economic land concessions.[3] Through economic land concessions,

---

[2] Social land concessions (SLCs) were created as a pro-poor policy to give land to landless or landpoor Cambodians in order to generate income from agriculture or establish residences. However, they have been strongly criticized by activists, land-rights groups, and communities as increasing land conflicts and deforestation. Conflicts have arisen for a number of reasons, including SLCs granting land that already had people living on it or being used by powerful actors to dispossess the land of the poor (Sovannara, 2018).

[3] Please see the website from the NGO Open Development Cambodia for an overview of economic land concessions and social land concessions, both of which have been implicated in Cambodia's land grabbing and deforestation phenomenon. As the NGO explains, an economic land concession (ELC) "is a long-term lease that allows a concessionaire to clear land in order to develop industrial-scale agriculture, and can be granted for various activities including large-scale plantations, raising animals and building factories to process agricultural products" (Open Development, 2015).

the government provides the land to the private company for planting agriculture, or like in the case of Think Biotech company, they plant acacia and eucalyptus trees. They then say they are replanting the forest, but in fact they clear the Prey Lang forest and plant trees for timber.

There is the collection of trees in economic land concession areas, but also from the protected areas, for example, by the two companies, Think Biotech and PNT.

Editors: Where do economic land concessions come from?

Prey Lang Community Activist: The government has a policy to provide land to private companies. Not only economic land concessions, but mining is also an issue, a part of the challenge for the forest, such as gold mining and also stone mining—clearing the forest to collect stones.

Editors: Who is profiting from it?

Prey Lang Community Activist: The companies in Cambodia, those companies, like Chinese, Vietnamese, and Cambodian companies . . . those companies that have good relationships with the government or with the high-ranking Cambodian officials. Only the corrupt companies come to invest in Cambodia! The relevant authorities, like the Ministry of Environment, do not think about the forest, they just try to get money, so there are many challenges for the forest.

Corruption is systematic, from the top, not from the ground. If it was from the ground, it would be easy to control. So how can we do anything about the corruption of government officials? If you want to get another license from the government, you have to give money to them, to high-ranking officials. So for everything you have to make a bribe.

Editors: Who is suffering from this?

Prey Lang Community Activist: Only the poor people. The poor people in Cambodia, those living and surrounding the forests. The poor people are crying every day, some people get arrested, some people lose their ability to work, and then they have to move to get new homes. After they change their home, what happens to them? The poor people are the ones that suffer.

Editors: Looking at examples across Asia and Africa, it seems that so many environmental problems are driven by a neoliberal development agenda. This includes increased emphasis on attracting foreign investors through many different investor-friendly rules and regulations, the collaboration between foreign investors and government officials, the emphasis on endless economic growth, and so on. Can you speak to this in Cambodia's case?

Prey Lang Community Activist: I don't know much about the case from other places, but in Cambodia, I think it is the same. But in Cambodia, the benefits fall only to individuals, it does not fall to the whole government. When they collect the natural resources from mining, from land grabbing, those things benefit the rich man, the government official. They do not benefit the whole government system. Cambodia is a poor country, so they need the natural resources and land for the development. But it is not for the people, it is for their own power. Not for the Cambodian people.

Editors: So the development just benefits the rich people or government officials? It's not benefiting everyone, the poor or even the whole government?

Prey Lang Community Activist: Yes, like Boeung Kak Lake. The lake used to be there, but the poor people surrounding Boeng Kak Lake got no benefit from those developments.[4] Like other companies, like PNT, the poor people affected by those companies don't get any benefit from them. For example, the authorities and the managers of the companies, they lobby the poor village people to buy the chainsaws to cut down the trees and sell the timber to them. But from day to day, they don't get much benefit. Some people, they take a microfinance loan to buy chainsaws or to buy the transport to sell the timber to the company. But I compared their situation with the local people who were not involved with the logging to see who is better. The poor people involved with the logging were not able to find the money to return to the microfinance,[5] because they did not benefit from their logging businesses. There are so many examples.

Editors: Who is providing the microfinance?

Prey Lang Community Activist: Small banks provide microfinance to get the benefit from the people. The companies go to people and lobby the people to take the loan from them, but the people don't have good knowledge about the loans. The companies go to the people and lobby them to take loans. They don't think about whether or not the poor people can pay back the loans. There is a lot of microfinance in Cambodia. Many people take microfinance loans. If people don't have the money to pay back, the banks take their houses and their land.

Editors: What is the hardest thing about your work in Prey Lang forest? How is it being an activist in Cambodia?

Prey Lang Community Activist: You know, the mindset of parents, when their son or daughter tries to be an activist, they always lobby them not to be an activist. They say "one day you will be killed or imprisoned," and it is the reality,

---

[4] For a full discussion of the Boeung Kak Lake evictions, see Chapter 6, "The Boeung Kak Lake Evictions and the Experience of Women in Cambodian Land Dispossessions."

[5] According to a number of recent reports by LICADHO, microfinancing has become deeply implicated in Cambodia's land grabbing through coerced land sales and also debt-driven hunger, child labor, and migration (LICADHO, 2021).

many activists have been arrested. They say "you are doing illegal activities," and the court charges them with one or two years in prison. Many are arrested without any evidence or reference to a specific law that says that these people did something wrong.

Editors: When you were working with PLCN, did you tell people you were an activist?

Prey Lang Community Activist: I always hid. Most people didn't know I was an activist. Going public, being an activist, is a lot of pressure. Government officials and authorities would have focused on me, making it very hard for me to work with the people. If we want to be a good activist, we have to hide. As an activist, I was not protecting my own land, my own forests. I was just the key for the poor people who were protecting their land and forests to be strong. I was like the gasoline, and the people were like the motorbike car. So I was never openly an activist.

Editors: All around the world, activists are fighting against deforestation and the land grabbing that usually causes it. Do you think there is anything different or unique about Cambodia's case?

Prey Lang Community Activist: You know, for activists in Cambodia, human rights activists, environmental activists, political party activists, opposition party activists, the situation is too bad in Cambodia . . . It's different from European countries, or Australia, or the United States. Cambodia is a democracy but political activists face too many challenges; they can easily be killed. Many people try to attack them. It is the worst for political activists.

Editors: Are you working with other activists in other countries?

Prey Lang Community Activist: I have some but not much connection with people not in Cambodia. In Cambodia, I have many connections with different sectors—political, human rights, environmental.

Editors: In your opinion, what are the solutions to the environmental/social issues happening now in Cambodia? What kinds of changes need to happen?

Prey Lang Community Activist: There is only one way we can have the solution. We need to change the government system in Cambodia. If we don't change the government, we don't have any solutions. The officials are deeply involved with environmental or social issues; most of those people are behind the issues. Many international NGOs helped to change the laws; there are many good things on the paper, but no one practices what is in those papers. The government must change.

Editors: How can you do that, change the government system?

Prey Lang Community Activist: Only by working with the grassroots people. If you try to work with the government, even if you provide them with so much money, they change nothing, because they want to keep their power. If they change, they will lose their power. We must work with the people, provide them training, tell them their rights; these are the ways to get solutions.

Editors: So you have to work outside of the system? Talking to politicians doesn't work?

Prey Lang Community Activist: Sometimes only international pressure from the big countries works. But the situation right now, the geopolitics, the conflict between China and United States, and Russia, is making it difficult. Sometimes the United States doesn't care about the actions of the Vietnamese or Cambodian government. They just lobby them to stay away from Chinese influence, and the Chinese, they also try to lobby the Cambodian government to follow them.

Editors: In Cambodia, are there many grassroots environmental activists? Is there an actual grassroots, bottom-up movement from the poor or those affected to change how things are happening? Have they been successful?

Prey Lang Community Activist: From my experience, I can say that many grassroot movements are here, but only PLCN is from the bottom up. One success of PLCN came in 2016, when PLCN convinced the Ministry of the Environment to make Prey Lang into a protected area.

Editors: Why do you think PLCN is so successful?

Prey Lang Community Activist: PLCN is the only independent grassroots movement in Cambodia who makes their own decisions. No one can put pressure on them to make the decisions. PLCN is like the stone—no one changes them, even me. Their heart is the love of the forest and want to protect it. So they don't change according to donors; they don't work for the money. But other grassroots movements are influenced by the donors. Many grassroots movements change their decisions because of their partners.

Editors: The people who fund you, you don't have to compromise, you can negotiate better with donors? Does PLCN have any donors?

Prey Lang Community Activist: They get some funds from donors, small funds provided every month. PLCN can negotiate with donors, and many donors that PLCN negotiated with are for the purpose of protecting the Prey Lang Forest. PLCN had only one donor who supported them during my role.

References

Flynn, G., P. Vantha and M. Erickson. 2021, April 1. "'What Other Country Would do this to its People?' Cambodian Land Grab Victims Seek Int'l Justice." *Mongabay.*

https://news.mongabay.com/2021/04/what-other-country-would-do-this-to-its
-people-cambodian-land-grab-victims-seek-intl-justice/

LICADHO. 2021. *Right to Relief: Indebted Land Communities in Cambodia Speak Out.*
LICADHO. Retrieved May 13, 2024. https://www.licadho-cambodia.org/articles
/20210628/173/index.html

Oakland Institute. 2014. "World Bank's Bad Business in Cambodia." https://www
.oaklandinstitute.org/world-banks-bad-business-cambodia

Open Development. 2015, November 1. *Concessions.* Open Development Cambodia.
Retrieved May 13, 2024. https://opendevelopmentcambodia.net/topics/concessions/

Prey Lang. n.d.a. Prey Lang Community Network (PLCN). Retrieved July 28, 2022.
https://preylang.net/about/the-forest/

PLCN. n.d.b. Prey Lang Community Network. Retrieved July 28, 2022. https://preylang
.net/about/plcn/

Sovannara, K. 2018, March 30. *Social Land Concessions.* Open Development Cambodia.
Retrieved May 13, 2024. https://opendevelopmentcambodia.net/topics/social-land
-concessions/

III

Pluriversal Perspectives for and beyond Resistance

8

Nourishing Communal Territories of Life

Indigenous and Afro-Indigenous Resurgences beyond Earth Crisis

Leonardo Figueroa Helland, Angela Martinez, Neftalí Reyes Mendez, Angélica Castro Rodríguez, Juan José López Negrete, Mileida Correa, José Gualinga, and Çaca Yvaire

Indigenous communality and reciprocity underpin processes to reconstitute biocultural territories of life. These processes embody decolonizing solutions to intersecting crises of health, environment/climate, food, and systemic state/corporate violence. Rooted in land-based communal self-determination, they revitalize Indigenous knowledge and governance. Upon Indigenous cosmovisions, they (re)weave sacred territories where biocultural diversity, food sovereignty, and political agroecology enact autonomous multispecies communities. Furthermore, these processes are committed to strengthening an intersectional community fabric that celebrates gender diversity and nurtures Indigenous-Black-peasant, people of color, and broader intersectional solidarities in struggles for systemic change. As opposed to market-based and state-sponsored colonial, patriarchal, capitalist, and neoliberal approaches advancing false solutions and techno-fixes, Indigenous resurgence processes move us toward territorially sovereign, anticolonial, anti-patriarchal, and noncapitalist forms of communal governance that actualize "pluriversal" worlds. We gather voices from four key processes from Abya Yala and Turtle Island ("the Americas") that embody decolonizing communal horizons of regenerative emancipation: (1) The Kichwa People of Sarayaku's Kawsak Sacha/Living Forest (Ecuador); (2) The Association of Indigenous and Afro-descendant Fishermen and Peasants for the Community Development of the Bajo Sinú (Colombia); (3) EDUCA, A.C., Services for an Alternative Education's campaign for Communal Alternatives in the Defense of Territory (Mexico); and (4) NEFOC, Northeast Farmers of Color Land Trust (Turtle Island/United States).

Weaving Communal Territories of Life Beyond Civilizational Crisis

An *Intercambio de Saberes* for a Tapestry of Resurgent Land-Based Cosmovisions and Emancipatory Regenerative Processes

This chapter is a tapestry of emancipatory communal processes, collectively woven by diverse locally situated voices, each grounded in nurturing rich biocultural territories that constitute actual horizons for regeneration, sustenance, and propagation of multispecies communities of life. Such processes are not "alternatives" to hegemonic orders but persistent (though subjugated) lifeways that precede, resist, survive, and will exceed a modern/colonial civilization in decay. Multispecies living territories regenerate via non-anthropocentric place-based, place-enabled, and place-creating bioculturally diverse communities that are the recursive outcome of reciprocally nurturing and cocreative relations among all species. Their territory as a whole is sovereign, living, and collectively agential, including land, air, water, other-than-humans, and the spirit, matter, and flow of all their relations. Human life forms are included but not exceptionalized as self-enclosed entities above others. Instead, human biosocial reproduction depends upon the collective determination of other-than-human agents/persons. All species exist not independently as objects or entities but as mutually constitutive and intersecting threads of codependent relations that make life's regeneration possible. These interrelated plural communities autopoietically constitute members as agents collectively and cocreatively reenacting the socio-ecologically reproductive and regenerative cycles that ensure continuity for healthy communities and territories for indefinite generations. Their cyclical and celebratory reenactment of life in its plentiful richness, diversity, and vitality is grounded in the sovereignty of the land and Mother Earth herself, the spiritual-epistemic acknowledgment of other-than-human personhood, cocreativity, intelligence and agency, the self-determination of communities, and the exaltation of a *vivir sabroso* (delicious living) whose beauty results only from the rekindling of communal energy or *Sumak Qamaña/Suma Kawsay/Lekil Kuxlejal*[1] (supreme energy-spirit-vitality cyclically regenerated by reciprocal conviviality among all community members).

All four processes below embody the (re)constitution of communal multispecies territories of self-determination upon cosmovisions of Indigenous,

---

[1] Comparable terms to Suma Qamaña (Aymara), Suma Kawsay (Quechua), and Lekil Kuxlejal (Maya Tsotsil) exist in other Indigenous languages.

Afro-descendant, Afro-Indigenous, peasant, fisherfolk, and forest peoples. These territories embody land, food, knowledge, and identity sovereignty and self-determination. They envision collective practices unfolding into horizons that, incommensurable and irreducible to global or universal logics, weave a pluriversal fabric of complementary lifeways. Such lifeways return, reconstitute, and recreate themselves to materialize actually existing worlds beyond an anthropocentric, patriarchal, colonial, capitalist, state-centric, and developmentalist model imposed upon a violent logic of global extraction. We center voices of these processes from across Abya Yala and Turtle Island to inter/bioculturally envision decolonial communal horizons of regenerative emancipation:

1) The Kichwa People of Sarayaku's Kawsak Sacha/Living Forest (Pastaza province, Ecuador). The Sarayaku people, also the Midday People, identify as Kichwa Indigenous with approximately 1,500 inhabitants in seven communal centers: Kali Kali, Sarayakillu, Chuntayaku, Shiwakucha, Puma, Kushillu Urku, and Mawka Llakta. They live in a richly biodiverse territory of 135,000 hectares including Sacha (forest), Yaku (rivers), waterfalls, black lagoons, Allpa (soil and subsoil), and Wayra (air). These sustain numerous ecosystems and species of flora and fauna, crucial to the livelihoods of families who manage chakras (native polyculture and agroforestry systems), harvest wild products, hunt, and fish. The Sarayaku territory is predominantly tropical Amazonian rainforest, a diverse landscape of hills, forests, floodplains, riparian forests, wetlands, salt licks, Mauritia palm swamps, and the Sisa Ñampí: "great path of flowers." The Sarayaku people have experienced pressures of colonialism, religious missions, rubber bosses, settlers, traders, speculators, police and armed forces, and extractivist corporate and state projects. Despite this, the Sarayaku people maintain traditional ways of ordering, using and managing territory, and organizing and relating with nature.

2) The Association of Indigenous and Afro-descendant Fishermen and Peasants for the Community Development of the Bajo Sinú (Colombia), ASPROCIG, emerges from the amphibious bioculture of the Sinú River's lower rivershed. An intercultural weaving process, ASPROCIG enacts a communitarian political organization among fisherfolk, Indigenous, peasant, and Afro-descendant peoples who share, defend, and struggle for the same territory. Institutionally co-governed by its associates, it conducts itself by an altogether *other* process, a development in its own

terms, outside and beyond modern/colonial/patriarchal hegemonic forms. Founded in 1991, it embodies an aesthetic proposal of development geared not toward accumulation, production, or consumption but toward nurturing the beauty of the whole community of life. ASPROCIG works at three scales: the Socio-ecologically Biodiverse Family Spaces (ABIF), the Collective Socio-ecological Systems (SSC), and Natural Ecosystems, including their revitalization of Zenú Large Scale Hydraulic Systems and Agroecological Systems in Diques Altos (SADAS). The aim is socio-ecological restoration at different community levels with a planning horizon of 200 years, well beyond the short-term vision of the dominant civilization, and toward a systemic transition beyond the dominant society. ASPROCIG gathers about 6,400 families (approximately 32,000 persons) in 74 communities.

3) EDUCA, A.C., Services for an Alternative Education (Mexico), is a Oaxaca-based nonprofit community organization. Founded in 1994, it advances communitarian democracy and development rooted in justice, equity, and participatory processes centering marginalized peoples. EDUCA works with Indigenous communities and organizations in the defense of territories and rights, particularly in predominantly Indigenous regions, fomenting participatory citizenship education, strengthening Indigenous authorities, communal knowledges, and alternatives. EDUCA strengthens the political role and presence of communitarian and social organizations by articulating civil society initiatives, territorial defense, and enhancing the visibility and influence of communities' political agenda. EDUCA's focus integrates community rights-based approaches toward deep democratic transformation. This includes the campaign for Communal Alternatives in the Defense of Territory.

4) NEFOC-LT: Northeast Farmers of Color (Turtle Island/United States) is a land trust gathering a hybrid community *and* conservation land trust model. It reimagines land access and conservation/stewardship of communities and ecosystems through a community vision that uplifts Indigenous, Black, and POC relationships with land, skills, and lifeways. NEFOC advances material and spiritual decolonization through the path of attaining permanent and secure land tenure for BIPOC through rematriating[2] land, seeds, and lifeways; farmland stewardship, preservation,

---

[2] Following Lenape and Shawnee scholar Steve Newcomb, rematriation entails restoring a people to their rightful place in sacred relationship with their ancestral land and Mother Earth (Newcomb,

and expansion; envisioning ways to be in reciprocity with land and creation; and reimagining what the word "farmer" stands for. NEFOC works toward the ecosystem-sovereign, BIPOC self-determining biocultural conservation of land through protecting native species ecosystems, engaging in food sovereign regenerative polycultural, agroecological and Indigenous farming and agroforestry, and advancing environmental policy that upholds the Rights of Nature (Personhood). It centers BIPOC voices and leadership and honors Indigenous sovereignty while healing colonial harm and protecting landed community futures by creating a carbon drawdown in the Northeast and informing climate policy locally, regionally, and nationally. NEFOC is sharing and advancing skills and knowledge with and for BIPOC farmers, land stewards, and Earth workers, connecting them with resources, training, education, and land to enhance their ability to thrive in ways that honor land and one another.

These processes have been gathered through ongoing conversations since early 2021, including a Mother Earth Day webinar.[3] The *intercambio de saberes* (wisdoms/knowledge exchange) proceeded in virtual form (partly due to the COVID-19 pandemic and geographical distance), including an exchange of writings and conversations leading to this collectively authored text. The facilitators, Leonardo Figueroa Helland and Angela Martinez, have sought to ensure that conversations and texts resulting from it are crafted upon the vision, cosmovision, knowledges, and guidance of the community leaders and organizers themselves. Among them, the voices and writers include Neftalí Reyes Mendez and Angélica Castro Rodríguez (Zapotec and Mixtec), Juan José López Negrete (Zenue and Afro-descendant) and Mileida Correa (Zenue), Angun [José] Gualinga (Kichwa of Sarayaku), and Çaca Yvaire (Atakapa Ishak).

Our intercambio de saberes weaves resurgent land-based cosmovisions and emancipatory regenerative processes as politics, theory, and method that upend and overturn hegemonic power relations of knowledge creation. These organizations advance epistemological frameworks collectively created, undertaken, and evaluated by their constitutive communities. This challenges the patriarchal coloniality of academic knowledge production and validation, which prioritizes specialized "expertise" legitimated through structures and procedures that generally embody Eurocentric, rationalistic, and highly

---

Steven. 1995. "Perspectives: Healing, Restoration, and Rematriation." *News & Notes*. Spring/Summer).
This must be done through depatriarchalization and thus recenters the reproductive and regenerative.
[3] Webinar: "Nourishing Communities of Life: Indigenous Resurgence beyond Syndemic Violence & Earth Crisis" (4/22/2021). URL: https://www.youtube.com/watch?v=pS5IsEUEBVc

institutionalized standards of research, science, scholarship, higher education, and authorship (the latter often highly individualized). Hegemonic politics of knowledge often reproduce, enact, and extend the dominant social-ecological power relations (anthropocentrism, patriarchy, coloniality, Eurocentrism, North-centric epistemic geopolitics, state-centrism, capitalism, etc.). Such power relations are the "social ecosystem" wherein dominant knowledge institutions operate and which enables their material-cultural reproduction. Hegemonic epistemic politics also frequently stratify and compartmentalize knowledge holders and knowledge forms, fragmenting and hierarchizing them through accreditation of individual scholars, scientists, specialists, and disciplines as legitimate experts and expertise. "Experts," to become and remain so, often must seek recognition/validation, accreditation, and livelihood—thus legitimation and material reproduction—primarily from academic institutions, which makes them fundamentally responsive to their institutional power and their political/economic/cultural enablers. In this hegemonic frame, knowledge is subject to legitimation upon its validation by standards, procedures, literatures, methods, and other highly ritualized processes largely embodied in the power of academic, scientific, and university establishments—and not primarily by the communities that embody the lived experience and knowledge that serves the reproduction of their societies and territories. These dominant knowledge institutions tend to be abstracted from the grounded conditions of territorial, intimately local relations of regeneration that are the basis of the biocultural diversity of life.

In challenging and subverting this hegemonic knowledge politics, this *intercambio* centers the cosmovision, voices, and practices of knowledge keepers and practitioners themselves: community weavers carrying embodied collective-territorial-community knowledge. The time is well overdue to recenter the community-grounded voices of those who have always nurtured and defended actually existing lifeways that do not reproduce themselves upon hegemonic political economies or their institutionally validated forms of power/knowledge. These processes thus constitute an altogether *other* epistemic politics, grounded in (re)affirming the territorial and authoritative knowledge sovereignty of communities and the cosmovision of peoples, autonomous and outside the universalizing logic of Eurocentric and patriarchal modernity/coloniality. These subaltern communities, via their collective *sentir-pensar* (feeling-knowing), help nourish the coevolutionary other-than-human—human biocultural diversity without which the cultural basis for reproducing biospheric life conditions is impossible.

The contemporary environmental-social crises manifest in the widespread destruction and dislocation of multispecies communities of life, the violent subjugation of territories to extractivist intervention, and the imminent revolt of peoples and Mother Earth against this violent and unsustainable project of dominion. These crises result, preponderantly, from the deliberate assault on the territories, cultures, and lifeways of Indigenous, Black, forest peoples, fisherfolk, smallholder, pastoralist, and peasant communities, carried out under (purportedly) civilizing, modernizing, and developmentalist visions. This continuing violence also carries the colonial imposition and globalization of an imperial Abrahamic (particularly, though not exclusively Christianizing) and Eurocentric heteropatriarchal order. These violences alienate regenerative land-based spiritualities/communities, favoring abstract projects of anthropocentric progress and mastery over nature consolidated in the, arguably "secularized," objectifying instrumental rationality and technoscientific managerialism of modernity. Real solutions to the epochal crisis of civilization cannot be within the same logic or by the same actors and structures causing, perpetuating, and profiting from its continuing violence. Only rematriating and restoring land, knowledge, food, and governance sovereignty for territories and local regenerative communities, rooted in communal and multispecies relations, can enact pluriversal pathways against and beyond the systems destroying life conditions for Earth and subjugated peoples.

Thus, the four processes speak in their own terms, upon their own authoritative epistemologies, from their own locations and through the rich living and spiritually grounded complexity of their knowledges, irreducible to each other or to any dominant logic of knowledge/power. Upon careful conversations, all agreed that each should speak from their own diverse text and location, irreducible to a universalizing logic that would impose a single narrative. First, each process writes below from their own positionality, knowledge, and vision. Then, upon the intercambio de saberes, we gather conversations to interlace their experiences, knowledges, visions, and pathways into a collectively authored conversation that reciprocally engenders key principles and pathways for the nurturance, defense, and reconstitution of a pluriversal tapestry of territories of life. The voices from these processes speak upon the authority and legitimacy of their own communal wisdoms and collective knowledges, rooted in intergenerational and relational communitarian processes of knowledge cultivation and sharing.

Weaving Communities of Life across Turtle Island and Abya Yala: Sarayaku, ASPROCIG, EDUCA, and NEFOC-LT

Sisa Ñampi—Frontier of Life; an Other Vision from the Original Peoples for Governance in Territories of Life, by The Kichwa People of Sarayaku, in the voice of Angun José Gualinga

TIAM, or "TRANSFORMATION," is the ethical-philosophical thought of the Kichwa of Sarayaku. It means an *other* vision, of return, of change, and of a new direction toward societal transformation—its necessity evidenced by the critical moments confronting humanity, like pandemics, Earth crisis, and the actual war among Northern, Western, and European powers.

Some years ago, in facing the aggression against the rights of Mother Earth, who is sick, injured, and endeavoring to heal, from the wisdom of the Sarayaku people, we proposed to the world the TIAM, calling for a reflection to build a new vision, a return to a life-centered and humane being. TIAM is a call to experience the Frontier of Life, the living Path of Flowers, *Sisa Ñampi*, a frontier that weaves and connects the existence of life in its fragile enactment between the limit of existence and death. This symbol, much more than a simple framing, roots the proposal and materialization of the Living Forest, Kawsak Sacha, a living being *subject of rights*.

This global proposal for an *other* paradigmatic vision calls for a life-change for humanity to return to the inner feeling of self with coexistence, which is nature herself. Thus, Frontier of Life, or *Sisa Ñampi*, convokes us to live in solidarity, share, and recover the path of the end and the beginning—return and rebirth. It is a symbol of life, of struggle and resistance to peacefully transform ourselves, care for land and Earth, and exist with our relatives of other cultures of the world.

Sisa Ñampi is to embellish and diversify the green rainforests of Pachamama so as to beautify with multicolor. *Sisa Ñampi* is to heal the soul and give happiness to the sad and afflicted. This titanic project has its roots in the struggle against oil extractivism, which symbolizes the limits of the life of the rainforest in the face of misnamed "development." It is a project that delimits the territory of Sarayaku through the beauty and symbol of a flower, opposed to that of a wall. Presently, this has advanced 110 kilometers of our territorial delineation to mark a limit to projects of extractivist death.

Sisa Ñampi was born from the great worldviews of the Living Forest, a lifeway of respect for the invisible beings we can nonetheless feel, who nourish and fertilize the Earth, the sacred mountains, lagoons, waterfalls, rivers, swamps, prairies, and trees, who are the dwellings of these beings where there exists abundance of fauna and flora, those whom we call Amazanga, Sacharuna, Yashingo, Yakuruna, and protective beings of the Kawsak Sacha and the Sumak Kawsay.

Beyond a local proposal, Sarayaku issued a global call for humanity to recover and return to feel Earth, so that the seas, glaciers, volcanoes, wetlands, and skies are recognized and declared alive and living. To recover what is lost, forgotten, is fundamental since it is not only the original peoples who keep this way of life alive. Other cultures have it too, but with the advancement of technology, industry, and capital, they have lost this essential life principle.

The original peoples have always, throughout history, through their legendary struggles and resistances, sustained the struggle for life and the rights of Pachamama. They have voiced historic declarations calling for the defense of Mother Earth, but lamentably, governments have not given them importance, and society remains deaf and blind.

TIAM proposes the rebirth, an *other* vision, that of the living being. That is the primordial understanding of the conception of life, as we are an embryo in the womb of Mother Nature, and only there will we respect and live tending harmonically to the resource that Pachamama gives us to live. Yet, today the opposite happens. Nature is seen as an other, an object of exploitation, leading us to disequilibrium and severe climate change. TIAM also proposes creating new indicators of wealth and value. These should measure wealth from a new and ecological vision; that is, the true richness and wealth that will keep Earth healthy and fertile with abundant fauna, rivers without pollution and abundant fish, life in solidarity, with sharing, unity, and equitable wealth distribution.

TIAM proposes radical changes in the world's educational system to incorporate this philosophy from families to educational centers, from earliest to highest levels, to achieve a real societal transformation. When human beings have recovered and accepted this original principle of respect to Pachamama and have felt the pain of the wound, then life will be reborn. TIAM counters the current consumerist and market-based model. TIAM is founded in the wisdom of the inter-dialogue of communication with Pachamama from which the Kawsak Sacha strengthens knowledges and builds paths of coexistence, connecting threads for the Good Living—Sumak Kawsay: the vision and path guiding our people's organization as a thread of life as a whole.

From this analysis, alternative visions emerge as proper solutions responding to ecological conflicts that original Indigenous peoples confront in facing alienating conceptions and worldviews that propose endless growth models, the (in)famous DEVELOPMENT that never achieves its final stage and instead continues its depredations against biodiversity and engenders the risk of collapse for life on Earth and the peoples' impoverishment.

In facing these crises, solutions emerge. These are rooted in millenarian knowledges and experiences of coexistence intrinsic to the Kawsak Sacha— Selva Viviente/Living Forest. These knowledges and experiences propose a new focus for organizing society to transform the present lifeway of depredation into one of tending respectfully for life's resources upon principles of conscious and responsible governance and the use of only what is necessary.

From the logic and wisdom of original peoples, in particular of Sarayaku, the rainforest is a living entity, who feels, has conscience, and is an active being. Thus, in this conception, we propose implementing the Sumak Kawsay model. Contrary to the word "Development," Sumak Kawsay values the elemental being of Mother Nature over and above a separated (abstracted) concept of a human being who is not primordial[4]. The notion of a primordial human has been inculcated in families and educational centers, leading to the logic that nonhuman nature is made to be exploited for the well-being of humans, which alienates us from the coexistence principle. This drives the direct and radical clash among the developmentalist Western logic/worldview and Indigenous Sumak Kawsay.

Sarayaku thus proposes its project of life through the Living Forest in its territory, embedding distinct key axes of Sumak Kawsay under the path of the Plan of Life, which are as follows:

Sumak Kawsay, as communal energy, gestates and enriches the exuberant vitality and lush diversity of life as a whole. Here, poverty (of people, land, or life) has no place as long as the living rainforest is collectively nurtured, cared-for, and protected. As long as it is unpolluted, flora and fauna will remain abundant and sustain community life for all. If the community nurtures rivers and forests so as to ensure rich abundant diversity of life upon a fertile soil, then collective happiness will be provisioned and life's regeneration will be sustained upon the wealth of values provided by the land. This is nurtured by communal organization

---

[4] That the human is "not primordial" means that it does not take on an ontological, epistemological, ecological, axiological, or political priority, superiority, or distinction (in the sense of power, exceptionalism, separateness, or privilege).

and kin relations as family, united by ties among land and community. Sumak Kawsay includes three pillars:

(a) *Sumak Allpa* (fertile land and soil). This requires organizing the community to ensure the maintenance of abundant fauna, conserved and enriched flora, unpolluted rivers with abundant fish, and agricultural practices with a diversity of products and cyclical regenerative rotation systems, among other practices. This encompasses all that refers to the Kawsak Sacha, the goods of life, inappropriately called "natural resources."

(b) *Runakuna Kawsay* (the people's life). This embodies the communal social life that enables us to exercise autonomous territorial governance under the people's self-determination. Upon this we build Life Plans (Planes de vida) for protecting, conserving, preserving and nurturing territory, and enriching biodiversity. This underpins visions for alternative economies/social ecologies that strengthen the grounds for collective happiness. This axis embeds the solidarity economy (including communal polycultures—chakras and forest gardens, planting, harvesting, gathering, fishing, hunting, artisanship, and the minga—communal labor). This undergirds communal authority and organization, including communal political, social, monetary, and basic services (e.g., communal social infrastructure, community trails, and community media).

(c) *Sacha Runa Yachay* (knowledges and wisdoms of the rainforest peoples). *Yachay* means knowledge for wisdom. It is a formative educational model enabling humans to live with/in the Ayllu (communal form of organization based on reciprocal balance and equilibrated collective labor). *Yachay* enables persons to achieve maximum knowledge and understanding of the Pluriverse of Kawsak Sacha in relation to the maximum respect for Mother Earth (Pacha Mama), which constitutes us as humans and carries through our wisdoms and knowledges transmitted historically and intergenerationally in communal and embodied-territorial forms. We articulate this in our cosmovision, bio-knowledges, culture, science and technology, health and medicinal practices, governance, and justice.

These pillars are embedded in *Life Plans*, cocreated and implemented communally as conducting paths that bring the Kawsak Sacha territory into existence as embodiment of the Sumak Kawsay. This is life in harmony for coexistence with the Sacha Runas (spirits, beings, and spirit beings of the forest) and to materialize happiness indicators of *Runakuna Kawsay* (communal social life). This Musky (vision) realizes goals of collective self-determination.

When we propose that the territory is Kawsak Sacha, we assume a new and different governance and co-governance alongside the Living Forest's protective beings. This is defined under free, autonomous, and sovereign self-determination and self-government precisely to conserve/nurture the richness and rights of Mother Earth. It is also defined through the tending and use of life's resources to engender justice in all forms (social, environmental, cultural, spiritual, and economic) toward the Sumak Kawsay of the people, in this case the Originary Kichwa People of Sarayaku.

To preserve and conserve the richness of biodiversity (the goods and resources of life, rivers, lagoons, mountains, fauna and flora, prairies, and wetlands), we organize territorial zonings and Indigenous mappings that strengthen and secure economic, ancestral, food, and knowledge sovereignty.

This governance that legitimates and recognizes the territories of the living beings with conscience is a pointed response against the institutionalized discrimination, extractive projects, and myth of the noble savage, all of which to this day impose laws without consultation, sharpening impoverishment and misery within the system and structures of the state.

In Sarayaku we understand from within the construction of a (re)newed society (upon the strengths of our knowledge which comes from the Living Forest and our ancestors, applying the millennial technologies to create models of solutions by the peoples, overcoming complaints of those who would reject the legitimacy of our cosmovisions, which we center to empower our communal wisdoms and practices within our territories) that no one will save the original Indigenous peoples; we will save ourselves.

Communitarian Wisdoms in Oaxaca: Collective Hopes for a More Just World, by EDUCA A. C. in the Voice of Angélica Castro Rodríguez and Neftalí Reyes Méndez (Services for an Alternative Education, Campaign for Communal Horizons/"Alternatives" in Defense of Territory)

In Oaxaca's communitarian history, alternatives have always existed and will continue to do so as long as the communities name, practice, and share them. From this stems the great importance of communicating their knowledges and wisdoms that envision, rethink, and carry out a return to what is theirs, to the root, to being and living in community—beyond the capitalist system.

The communities conceive of their lifeways in resistance as ever-present "systemic alternatives" whose movements embody the reproduction of ancestral communalities into the future. These communities and movements form webs that define alternatives as *already existing* "pathways and horizons" central and common to their societies, built historically, which peoples and communities transit daily. These alternatives proceed from a collective commitment organized to reclaim and revindicate our existence, satisfying our basic necessities, and pushing proposals toward social change and utopias of constructing hopes that are only possible upon comunalidad (communality).

To think of other forms of interrelating and to resignify how we organize, defend, and protect territories and share our knowledges are fundamental practices in this moment. One example is the actions of the Front Against Mining for a Future for Everybody (Frente No a la Minería por un Futuro de Todas y Todos), a network of communities organized into a social movement in the Central Valleys of Oaxaca. Starting in 2009, people there have experienced the impacts of the mining project "San Jose." The mining project compromised the social fabric and safety of the community of San Jose del Progreso. Two land defenders were assassinated. In the face of this violent situation, the Front rewove and reconstructed organizational forms and internal systems to strengthen the social fabric and bring peace in the region. Their efforts started with installing Movimiento Radio ("Radio Movement"), a youth collective concerned with sharing their ideas and publicizing real information rooted in the feelings of their community. They also installed community gardens. Further, the region's communities created information-based community assemblies, resulting in the rejection of any and all extractive projects in addition to promoting cultural and political actions to defend the territory.

Another action has been the "Peoples' Guelaguetza[5] Against Mining." In said festivity, there is conviviality among communities, taking cultural expressions like dance and music to reclaim and revindicate Indigenous Zapotec identity, and engage in non-monetary and complementary community exchange of goods. Maize, squash, mezcal, flowers, seeds, and legumes are shared among everybody toward the end of the festival. This people's Guelaguetza embodies the "give and share" and the "enjoyment-gaiety" enacted by Indigenous communities in

---

[5] Guetza (Zapotec) or da'an (Mixtec) means mutual help. Traditionally, Guela-guetza refers to celebratory or festive communal mutual help. The term "guelaguetza" is today contested since the state government and private sector have appropriated it for events and spectacles that do not embody and may counter the interest of many Indigenous and grassroots communities. Hence *peoples'* guelaguetza's explicitly counter-hegemonic appropriations.

their relation to territory. From the start of the festival, there is an offering to Mother Earth of what is produced through collective labor (tequitl/tequio): tortilla, maize, water, flowers, and candles. This festivity is framed in relation to the Statewide Day of Rebellion Against Mining in Oaxaca, declared through an agreement made in the community of San Jose de las Huertas in 2017. That event saw the participation of movements and communities in resistance against mining across Oaxaca, because in each community there emerged a commitment to revindicate and reclaim the right to territory, free self-determination, and dignified life on this day.

In confronting an economic system that deepens inequality, inequity, dispossession, and violence, Oaxaca's communities nurture pathways toward a just and dignified life. In their words, "it is necessary in these times, to return to what is ours, the commons, to the root, the land, Earth, and the collective." The present global crisis prompts urgency to rethink ourselves collectively and to reclaim and revindicate our contributions as peoples, our resistances and struggles, our proposals, and our social and political visions, translated into communitarian organization. Such organization is accompanied by our communitarian knowledges. The collective/communal work (tequitl/tequio/ guetza/da'an), traditional fishing, food sovereignty and self-sufficiency, the communal fiesta, the communal assembly life, the systems of communitarian organization, the internal and customary normative systems, and the spirituality rooted in Indigenous land-based cosmovision embody the key pillars to live in community and build possible and real hopes rooted in practice and quotidian life. Territory and community are spaces in dispute where we reclaim and reproduce the community of life translated into social, cultural, political, and economic proposals. Rooted in Oaxaca in relational webs with other territories, there exist diverse efforts to defend this communitarian lifeway practiced daily. It is important to reclaim such efforts to build a world with social justice. We deem this the horizon to build collectively.

A Disruptive Process in Search of a Just and Sustainable Society: ASPROCIG, by ASPROCIG's Support Team and the Directive Committee, in the Voice of Juan José López Negrete

The Association of Fisherfolk, Peasants, Indigenous and Afro-descendant Peoples for the Communitarian Development of the Great Marsh of the Lower Sinú (*Asociación de Pescadores, Campesinos, Indígenas y Afrodescendientes para el Desarrollo Comunitario de la Ciénaga Grande del Bajo Sinú, ASPROCIG*) is a

Grassroots Communitarian Organization (Organización Comunitarias de Base (OCB)). It is constituted as a network integrated by ninety-eight Grassroots Communitarian Organizations across nine municipalities of the lower basin of the Sinú river. It constitutes its territoriality in the northwest of the Colombian Caribbean. ASPROCIG gathers 6,400 families (approximately 32,000 persons) in 74 communities. It is directed, administered, and dynamized by its own affiliates and constituencies across three spheres: the General Assembly, the Directive Committee, and the Support Team.

The organization was founded on February 2, 1991. Its objectives are the defense and struggle for land and territory, food sovereignty and security, and poverty elimination. Its present focus is on a gradual transition toward a just and sustainable society that embodies the aforementioned objectives. We have thus formulated a Proposal for Socio-ecologically Focused Territorial Development (*Propuesta de Desarrollo Territorial con enfoque Socioecológico (PDTeS)*). PDTeS, which is rooted in a paradigm of intercommunal human and other-than-human complex relationality, constitutes territory as an integrated unity, abandoning the urban/rural dichotomy. We understand territory as an indivisible whole wherein humans with their social structures and cultural constructs interrelate with the rest of nature's members.

The PDTeS has ten programs, collectively adjusted every five years. Presently, these are (1) food sovereignty/security, (2) education (Socio-ecological Peace School), (3) gender and intergenerational equity, (4) climate change adaptation, (5) solidary exchange system and commercialization, (6) potable water and basic sanitation, (7) restructuration of Natural Ecosystems, (8) communitarian tourism, (9) political incidence, and (10) institutional strengthening.

To address the present objective of transitioning toward a new society, we endeavor to supersede positivism as the philosophical paradigm for interpreting and organizing reality and move toward the relational complexity paradigm. This entails undoing forms of interaction and organization among "humans" and with "nature" based on separated and objectified ontologies and categories. This paradigm also seeks to supersede the rationality of human separateness, alienation, and domination among humans and with other-than-humans and living territories. Furthermore, it works to enact mutually regenerative communities of life wherein territoriality is constituted dynamically by the complex and reflexive relationality among interrelated and reciprocally constituted webs of human and other-than-human spirited agents. Respect, solidarity, love, and friendship among all relations are the guiding values among all members in dynamically enacting these communal territories as organs of life as a whole. Doubtless, this is a new field to imagine and implement development beyond the mysticism, ideologies, religious conceptions, party politics, dogmas, and institutional and organizational forms built upon the positivist sciences. ASPROCIG uses the letter Z to convey its methodology (see Figure 8.1). The two horizontal components of the methodology are (1) local knowledges, understood in the framework of pluriculturality dynamized by permanent dialogue and as a property of biodiversity, not only as a human cultural product; and (2) the "doing," as the main form of learning, dreaming, creating, concretizing, and feedback. Finally, the transversal connector of relational and communal values permeates the whole process: the feelings, especially love, friendship, honoring, respect and solidarity. ASPROCIG's Methodology Z can be synthesized as in Figure 8.1.

It is through our Methodology Z that we advance the PDTeS. We endeavor thus to transition to a just and sustainable society which would surpass the main problems humanity confronts today. The socio-ecological focus generates a conceptual framework, orienting a dynamic process where a living, intercommunally enacted biocultural territory is the unfolding spatiotemporal unit for relational action and collective transformation. Development from this perspective no longer aligns with the dominant logic of production and consumption. Development must be neither universal nor linear, or based on

Local knowledges, understood in the
framework of pluriculturality
dynamized by permanent dialogue
and as a property of biodiversity, not
only as a human cultural product

The "doing," as main form of
learning, dreaming, creating,
concretizing and feedback.

Figure 8.1 Re-elaboration by the authors inspired on the diagram produced by ASPROCIG and as published in Gonzales-Madera 2021.

mastery of humans or other-than-human nature. There is not one "development" nor an objectified or teleological positivist rationality to its "progress" (whether as modernization, developmentalism, or linear evolutionary, technology-centric, and/or reductionist science-centric progress). Contrastingly, there is a plurality of developments unfolding in different directions and enacting worlds of aesthetically substantial difference. This rich diversity of plural pathways is constructed not according to a universal positivist science or linear course or direction but as an aesthetic collective endeavor seeking to differently beautify the experience of life as a whole. Proceeding from Afro-descendant, Indigenous, peasant, and fisherfolk lifeways, the aesthetic focus of enacting territorial communities is to enable "vivir sabroso" (delicious life and delicious living—a collective experience integrating humans and other-than-humans). This aesthetic of plural forms of vivir sabroso cocreatively weaves territories in relational community while disrupting and subverting the claims to universality of the dominant positivist rationality. The guiding purpose is to nurture enriched places and life-webs where living becomes a delicious experience of collective and relational coexistence, cocreation, and coevolution with the totality of nature. This approach understands that any "civilizatory" human project is necessarily an indivisible part of a collective human *and* other-than-human craft of cocreating nature in emphatically diverse ways. This spirit carries a distinct epistemological proposal counterpoised to that which governs Western society. The latter is based on considering humanity outside of nature. Further, it considers nature as an instrument for human development to the absurd extreme of commodifying nature without considering its consequences.

This conceptual framework is carried out via permanent feedback processes among ASPROCIG's constitutive communities. It is dedicated to designing reciprocally supported spatiotemporal territorial units at distinct scales which are nonetheless autonomous. They mutually complement each other yet operate as fractal structures conserving their basic properties (e.g., self-sameness, complexity, and dimensionality). They have three structures. First are the *ABIF*, the *Socioecologically Biodiverse Family Spaces*, oriented toward coexistence and coevolution toward "Vivir Sabroso." We consider basic fractal units which in their proper grounded conditions give origin on their own to Socio-ecological Communities. These are considered the keystone of the process of transit toward the desired ideal of society: zero-hunger, balance of powers, gender and intergenerational equity, local capacity for the resolution of conflicts (peace), biodiversity conservation, decentralized energy systems with low GHG emissions, local economy based on principles of subsidiarity, equity and solidarity, and other components of sustainability. Second are *Collective Socioecological Systems (SSC)* integrated by family groups ranging from 15 to 80 families and areas between 5 and 15 hectares. Their principal objective is strengthening the cultural subject, meaning communities and territories that embody the autonomous identity, cosmovision, knowledges, practices, and aesthetic collective vision relationally co-nurtured by fisherfolk, Indigenous, peasant, and Afro-descendant communities. By virtue of this objective, families experientially learn deliberative and participatory democracy, conflict resolution, restorative justice systems, collective leadership, and balance of powers. Third are *Natural Ecosystems* wherein distinct communities, depending on defined Natural Ecosystems, coordinate to achieve biocultural restoration and sustainable management. This involves the distinct exercising of power and governance integrated into said ecosystems. These processes enact the collective sovereignty of territorial communities in all forms: knowledge, governance, land, food, and ecosystem sovereignty. To exemplify, ASPROCIG is revitalizing Zenú Large Scale Hydraulic Systems, SADAS, not as exclusively Indigenous but through cocreative collective innovations among fisherfolk, Indigenous, peasant, and Afro-descendant communities.

ASPROCIG's timescale for materializing its goals is 200 years. It thereby distances itself from present trends set by the logical frameworks derived from theories of change and the subsequent results-based methodology, hyper-saturated by measurable and quantifiable indicators. Methodology Z, designed to operationalize the PDTeS, is based on processes featuring complexity indicators. Separate from the Cartesian rationality of objectivity, these processes are rooted in modes of organization denominated as "Consensus of Subjectivities." This is a way of revaluing what is felt while setting aside the rationality of measuring or quantifying. From that temporal horizon, we have moved thirty years. At this stage, the work with children and youth is an existential priority for the organization.

"Making Worlds Is Not Limited to Humans," by Northeast Farmers of Color Land Trust, in the Voice of Çaca Yvaire

Making Worlds is not limited to humans

—Anna Lowenhaupt Tsing

Northeast Farmers of Color Land Trust exists as an in-between point where plant and animal nurturers and other people of the global majority intent on liberation are declaring spaces for alternatives.

Ours will be an artifact of communal reciprocity parallel to, though separate and withdrawn from, the settler fiction of the United States. We have no interest in projects that reinforce the false settler fiction of the United States because we cannot achieve what our ancestors labored for if we seek to produce environmental, social, psychological, medical, educational, and artistic forms that do not upend the human versus nonhuman binary present in Eurocentric and settler administrative practice.

Our work is akin to setting up a sanctuary for devotion to the shapeshifting methods that have allowed our collective and diverse peoples to navigate the contractions of a Eurocentric civilizing perspective. We are culturally remaking the roads along the rivers, inciting provocations for an immeasurable, more holistic study. We are workers who know that the beings in a given place—a land, a territory, an ecosystem, a community of life—are our family, our relations, with whom we must nurture intimate interpersonal belonging.

We are organic beings with bounded relationships with the nonorganic—what is produced—from which we must emancipate ourselves. We gaze at the stars. We have amnesia and we remember. We live through crises. We have always lived through crises, especially as Indigenous, Black, and people of color. We live through crises by practicing an embodiment of the futures that was sung around matters of cavernous, powerful concern—our "center does not hold, it stretches," to paraphrase Chickasaw scholar Jodi Byrd.[6] Our record is one of building. We are not outliers nor are we under any illusions that we have infinite control or authority over the ethics, material, or congressional, as they correspond to our rematriating practice. But we do see the power in diverse peoplehoods and recognize the need for a new system of jurisprudence, one that allows native and nonnative nations the opportunity to deepen the growth of their soul, cultural, and individual; a jurisprudence that presents a lantern for the world outside. There are forces in motion that we cannot meet well if we choose not to meet them together. For that, we need trust, reciprocity, and consent in our interactions.

That is where our land trust sits—as space to build greater trust in human interactions. We are political in that we wish to reorganize the means by which power is recognized, practiced, and released. We are democratic in the sense that we too wish to be released from the constant deliberation regarding what violent encounters are acceptable and which are not. This means that we work to emancipate our communities and all our relations (human and other-than-human) from any and all forms of social compulsion impelling us to deliberate

---

[6] Byrd, Jodi A. 2011. *The Transit of Empire*. University of Minnesota Press.

what forms of violence we are willing to reenact and be complicit with in order to secure our survivance into a future that would not be of our cocreation. We seek total liberation from compelled futures not envisioned by us; we seek futures rooted in the sovereignty of ecosystems. We endeavor to rematriate stolen territories to the other-than-humans and humans who have nurtured them. We endeavor to pick a path that can be moved along with the intent to shield our peoples from the capitalistic and colonial degradation of dominant cultural ways. And so we make propositions grounded in our cosmovisions with all due consideration and a healthy respect for contradiction. This is because we understand that our communities and relations are presently constrained, and must persist and exist under conditions imposed upon them, not of their own creation. Yet, we are committed to full self-determination, to the cocreation and regeneration of territorial reorganization in accordance with the vision of sovereign territories grounding BIPOC worlds and decolonial futures. Self-determination is imminent and we will be successful.

We employ experimental forms of administration designed to bend legislation toward multispecies jurisprudence in which we employ a critical Indigenous pedagogy woven through conversation with many different Indigenous epistemologies seeking to step into a peoplehood that is planetary. Crafting the space in which we can gather to make decisions together regarding how we will occupy this moment in these systemic crises is just an opening. The Sovereign Ecosystem Trust is a project initiated through the Community Conservation at NEFOC in relationship with Earth Law Center attorneys. The Sovereign Ecosystem Trust is beyond both conservation and community land trusts, encompassing all forms of Life-Mantle of the Earth entanglements in order to initiate the reclamation and rematriation of lands, and enable land access, control, and sovereignty to BIPOC land tenders with a view to sustaining and remembering our relationship within diverse kin-wise webs of life. Using the Sovereign Ecosystem Trust, we cannot hold ourselves to a standard that is not grounded in the community of human and other-than-human relations that presents itself in the places we find ourselves inhabiting. This is why we reject state-determined borders and denounce capital alienation.

Interweaving Emancipatory Projects: Wisdom-Sharing, Knowledge Cocreation, and Strategic Covisioning

This section is an interpretive discussion/reflection upon conversations among the above processes and text authors. While writing has been interwoven by Leonardo

Figueroa and Angela Martinez, the content is coauthored from the words and collective analysis of all contributors in this intercambio de saberes.

As Angélica and Neftalí from EDUCA underline, we must proceed from reaffirming that the so-called "alternatives" have always already existed, embodied by communitarian and communal lifeways, ontologically distinct from the hegemonic socio-ecological, political, and economic perspectives. For the communities, their lifeways and visions are not "alternatives" but their own main ways. Moreover, they are not isolated but interwoven into collective processes. The voices foregrounded in this intercambio among Sarayaku, ASPROCIG, NEFOC, and EDUCA demonstrate that distinctive local communitarian processes share communal orientations and values, opening spaces for interlacing their struggles. While politically and economically distinctive and unique, grown out of their biocultural and historical contexts, they also articulate via multi-sited struggles for land-based justice and equity and are thus bases from which we can nurture a different politics. Important across them is the recognition that their communities are not inventing new things. Rather, they are lifeways revitalized, reaffirmed, refigured, defended, and resiliently sustained, having always already existed. They are not isolated alternatives, nor exclusively economic or human, and do "the political" differently in centering multispecies and intercultural conviviality and cocreativity. This counters hegemonic ways of doing "the political," which are rooted in human-centric (particularly male and Western/Westernized) legal and political personhood, representation and governance rationality, and state/capital-centric property and governance regimes. These processes enact worlds that converge toward the communal, collective, and relational beyond the individual and are rooted in quotidian practice with others and the land. They are not to be subsumed to abstract generalizations and logics, nor reduced through equations to other struggles (Angelica and Neftali, EDUCA). Moreover, as Angun from the Kichwa People of Sarayaku notes, their visions reflect a common underlying force that attracts different, locally rooted endeavors toward a shared orientation, namely that of nourishing and beautifying the "common house"—the living community of life.

Furthermore, Juan Jose López Negrete, Eber Grondona, and Mileida Correa of ASPROCIG emphasize that the perspectives of all four processes, embodied in the writings, denote the peripheral. Yet the "peripheral" is not the same as that distant from the center. It is, instead, that which *refuses to claim the center and roots itself always in the intimate commitment to the health, conviviality, and liberation of the local.* Each of these processes embodies the conceptual periphery grounded in non-universalizing, nondominant, nonhuman-centric as well as subaltern human knowledge. Such knowledge embodies the structural axis of multiple other (path)ways of "development" or rather complex webs woven out of myriad possible multidrectional cocreative pathways, different from the positivist universal teleology of modern/colonial development. They are thus peripheral not just in the sense that they are rooted in the axis of the local but in the sense that they are rooted in different modes of knowing that are never reducible to universal knowledge and can only be understood through intimate relationship nurturance, cocreation, and conviviality with each unique community of life, including diverse other-than-humans, land/waterscapes, and humans. These processes thus assert that there is no such logic as the "global" and endeavor to refuse, exit, and escape the presumptions of globalization and of the global that would make everything translatable, replicable, and tradeable in relation to its totalizing language, logic, law, and economic-cultural currency. Only the local-intimate exists, and all else is epiphenomenal, since it is from local ground and in translocal webs with other localities that all material life is nourished and regenerated. That is why, when we interweave our processes, labor, and writing, we do it situated in the spatiotemporality of the local, wherein the development of one process makes sense only without overrunning the others. Furthermore, it is also important to acknowledge that each process is situated in different power relations and struggles in their own context and should not be universally reduced to a struggle that can be generally understood. Instead, knowledge has to be *intimated* through relational nurturing: nurturing relations upon which to feel-know-cocreate each other among local communities— including other-than-humans and humans. The respect for the irreducible difference of each collective voice brings separate texts in translocal interlacing, which is why the texts must be presented separately at the outset of this chapter. This proceeds from a recognition of each territory as a basis for qualitatively and ontologically distinct worlds of meaning-making in collective land-rooted practice (Negrete, Grondona, and Correa, ASPROCIG).

As Juan Jose López Negrete, Eber Grondona, and Mileida Correa of ASPROCIG articulate, this heuristic embodies, for example, the ASPROCIG process itself, wherein Afro-descendant, fisherfolk, Indigenous, and peasant peoples gather in shared spaces but do so proceeding from a deep respect for the differentiated identities that accompany each other and walk together in common pathways. Thus, the guiding question orienting us is how do we establish right relationships and respect among all diverse communities and beings, other-than-human and human? Moreover, how do we establish shared paths without overrunning each other? We do not create a sole organizational logic but proceed from the recognition of territory as a homemade space. For example, 80 percent of the territory home to ASPROCIG communities is watery (wetlands, estuaries, coasts, lakes, lagoons, and seascapes). Therefore, the communities in that biocultural region have nurtured an amphibious bioculture, a culture among living water, shaping themselves to live and develop by what the living territories prescribe (as Zenú Indigenous peoples of the region have always done). These communities exist among land and water and move in between, from land to seas to estuaries and watersheds. This is easier for communities to practice, than to speak or write about, since such Afro-descendant, fisherfolk, Indigenous, and peasant peoples are a spoken living culture. Text, even when interwoven by a continuously nurtured relational intimacy, can only temporarily crystallize in word what is a continuously regenerative cocreative communal act of living in movement, like the ancestral and contemporary amphibian bioculture of moving between water and land. Our lifeway learns to exist in accordance with the amphibious movement prescribed by the living territories.

As conveyed by the ASPROCIG team, these are, however, intervened and disturbed territories, where nothing is pristine. On the one hand, Indigenous and other peoples have cocreated and co-nurtured the richness of life here since time immemorial. On the other, third parties have sought to violently appropriate territories into a macro-project; a violent global logic of domination and extraction, reducing, homogenizing, and subsuming the unique personhood of each living community. In fact, all processes gathering in this space, from South to Meso to North America, have been local communities of life subjugated and intervened in the service of such macro-projects that sacrifice the local life to a global teleology of positivist development. And yet, we are healing these territories, as they are the common home and house of us all. For example, the Cauca (southwest Colombia) is an intervened territory that has been the site of divisive state strategies fostering confrontations among Indigenous and peasant communities and advancing extractivist projects. All our processes have sought to build bridges among identities without dissolving their distinctiveness. Restoring beauty in common life is key, whereby territories subject to extractive projects of death (e.g., exploited watersheds, forests, marine and coastal areas impacted by hydropower and megadams) give way to cocreated territories of life. These projects of death see the world as an object of consumption, and within that logic there are no real solutions.

The transition must be led by local ecological knowledge, interculturally weaving diverse biocultures. Similarly, this must be premised on the acknowledgment and primacy of the knowledge embodied and enacted by nonhuman persons like plants and broader ecosystems. Furthermore, reconstituting and interweaving subjugated knowledges like Indigenous, Afro-descendant, fisherfolk, and peasant knowledges with critically oriented knowledge of Asian and European arrivants is key to an intercultural sharing of wisdoms for the nourishment of a new society. But this requires that no knowledge system overpowers others. We must nourish shared epistemologies rooted in the common house, the whole community of life, beyond decolonization. This requires centering local knowledge not exclusively in the anthropocentric sense but based on the recognition that plants, animals, and ecosystems have their own knowledge. It is the local knowledge of the community of life that must share the broader common house.

The dominant positivism of modern/colonial civilization has given way to a complexity science where there is an opening to recognize local knowledges. However, this must recognize both Indigenous and local, as well as nonhuman knowledge and agency. For example, ASPROCIG, which seeks to embody a conceptual model that can be shared with other local endeavors, includes work of intercultural knowledge sharing in its socio-ecological school for peace, which has been active for nine years. Powerful groups have been undermining hope and trying to erase the fact of other lifeways and a vision of a different history and future. For example, colonialism and modernization have sought to homogenize societies within and under the Eurocentric world-system while invisibilizing the continuity of non-Eurocentric modes of creating/organizing social ecologies and knowledges (e.g., Indigenous knowledges). The human brain and spirit are also a territory in dispute. This is primarily in the struggle for recovering knowledges and territories where another way of life is possible and where motivation can take root into actions that sustain and create this other way of life. Each of the processes in conversation attests that there is not just one hope but many hopes flowering into a great pluriversity of distinct, interlaced worlds that can reconstitute life's fabric (ASPROCIG, Juan Jose López Negrete, Eber Grondona, and Mileida Correa).

With regard to the biocultural and political context of Oaxaca, Angelica and Neftali from EDUCA underline that in Mexico there exist fifty-six recognized Indigenous peoples, with Oaxaca boasting the largest diversity of sixteen Indigenous nationalities, plus an Afro-Oaxacan population. In certain parts, Indigenous and Afro-descendants live in conviviality. For example, the Mixtecs and Chatinos live in the Rio Verde (Green River) region with Afro-descendants. And while Indigenous people's rights and recently Afro-descendant rights are recognized in the Mexican state's federal constitution, as those of other diverse identities, the fact is that living territories are similarly intervened and assaulted by companies and "development"—including "green development"—projects. Yet, communities embody existence as resistance rooted in territorially embedded knowledges, wisdoms, histories, and lifeways which have created some of the richest biological and cultural diversity (biocultural diversity) anywhere, coupled with one of the areas with the highest density and plurality of self-governed/autonomous Indigenous, peasant, fisherfolk, and Afro-descendant territories across the world. Such communities that exist in their own long and complex living histories of land-embedded knowledge and wisdom are targeted by government, corporations, and dominant social groups (e.g., settler, metropolitan, technocratic, developmental, or scientific agents) on the pretext of development and now "green development" (e.g., hydropower, agro/biofuels, industrial wind, solar and geothermal plants, technology minerals mining, forest, carbon and biodiversity markets, and enclosures and offsetting schemes).

Communities are constantly portrayed as "anti-development" and asked by dominant groups to not just "oppose development" and "resist progress" but to propose some "alternative." But their lifeways have been the basis of living territories for thousands of years before "development" arrived. They are not "alternatives" to development but indeed precede development. Further, how they have tended to lands cocreates and maintains the so-called "natural wealth" that landed castes, corporations, and states have sought to exploit and without which they would know of no such resources to control and accumulate. Therefore, grassroots communities, collectives, and organizations like the *Colectivo Oaxaqueño en Defensa de los Territorios* and *Servicios para una Educación Alternativa* work upon the collective strength of communitarian organizing and Indigenous, peasant, fisherfolk, and Afrodecendant cosmovisions to defend territories against extractive projects. This is an anticolonial and anti-capitalist struggle for a dignified life—a struggle not to create new alternatives but to defend/revitalize already existing ones that have been here, evolving and recreating themselves as the communitarian basis of life for thousands of years. This already underlines key convergences with other processes in this sharing space, like Kawsak Sacha, which are based on the sovereignty of territories and ancestral communities to define priorities according to what is needed to regenerate life and not what serves a logic that appropriates territories for distant projects of extraction (Angelica, EDUCA).

Angun Gualinga explains that in the Amazonian territory of the Kichwa People of Sarayaku, in Ecuador's Pastaza Province, Indigenous peoples have co-nurtured with nonhuman spirits and persons a very well cared-for and conserved territory: a deep rainforest of 135 hectares with rich biocultural diversity featuring black lagoons, rivers, and waterfalls that run downstream to Peru. There are ample landscapes which Indigenous nations share, such as the Waorani, Achiar, Shuar, Sapara, and Andua. In the Amazon region of Pastaza alone, there are seven Indigenous nationalities, each with caring and conserving living territories. Deep in the rainforest, there are over 130,000 hectares tended by Indigenous peoples; 95 percent of these are rich in biodiversity. These areas contain a wealth and variety of different environments, climates, and species with which Indigenous communities have complemented each other for thousands of years. Legally, recognition of collective Indigenous title exists since 1992. However, the state embodies strategies of intervened territories, subsuming them into a macro-extractivist logic, as it continues to divide the living territories into blocs (e.g., for fossil fuel extraction), producing various conflicts. The Amazon community of life has suffered the imposition of state-sponsored colonial and capitalist geography—a map of extraction and concessions. For example, the Kichwa People of Sarayaku have been fighting for four decades against oil extraction. Neocolonialism is encrusted in the state's governance structures. Modern states like Ecuador may legally recognize plurinationality, but it is peoples and communities that must (re)build lifeways that embody the complex state of plurinationality beyond a dominant system.

As highlighted by Angun Gualinga, all processes in this knowledge-sharing space embody communities that confront entrenched neocolonial institutions and are working toward autonomous decentralized governments in accordance with territorialities dictated by the spirituality and sovereignty of the land and honoring Indigenous cosmovision. Colonial, state, and corporate geographies create and impose territorial plans that exclude communities. The legal recognition of Indigenous and racially subjugated communities by states may seem inclusionary. However, in practice, this often sacrifices their territorial visions *beyond* the state under a larger project of modern development, including eco-modernization, which perpetuates the eco-genocidal and colonial logic of extraction.[7] Indigenous peoples often see the state of their communities in the law and the law of the state as a literal dead letter as it erases and kills our ways of life and the life and spirit of our territories and nonhuman companions. And yet, Indigenous peoples have sought to resolve conflicts on their own terms,

---

[7] For detailed critical discussions and numerous sources in support of this statement, see Figueroa Helland, L. E. (2022). "Indigenous Pathways beyond the 'Anthropocene'". *NYU Environmental Law Journal (ELJ)*—Special Volume (30) on the 2022 Symposium: Free the Land—Land Tenure and Stewardship Reimagined. Available at (nyuelj.org).

like through the Confederation of Indigenous Peoples of Ecuador (CONAIE) (Angun [José] Gualinga, Kichwa People of Sarayaku).

At a broader and deeper level, centering a more complex historical, cyclical and multispecies horizon beyond Eurocentric anthropocentric politics and based on Indigenous cosmovision and spirituality, Angun Gualinga notes that when we speak of plurinationality from below, from the communities, it also means the recognition of nations of other-than-human persons who make the big house, the community of life, possible. That is why the Kichwa People of Sarayaku propose governance rooted in Kawsak Sacha, the living (forest) community of life. This merges cosmovison and politics (we could say an *other* cosmo-politics) wherein territories are recognized as living subjects entitled to rights. This could be seen as similar to the proposal by NEFOC companions for a living sovereign ecosystems as the basis of autonomous, non-state-based Indigenous communal institutionality and jurisprudence. The exercise of governance from the sovereignty of the living territories is an autonomous exemplar that is not based on exclusionary "protected areas," as is common in Western objectifying and positivist logic, and makes no sense to Indigenous communities. The Sarayaku proposal is one of co-governance along with the Living Forest and all its beings, which is rooted in the acknowledgment, recognition and respect of the spirit, personhood and agency of the forest, and the spirit of all its beings as protectors of the community of life. The Kichwa People of Sarayaku propose the Living Forest on the basis of a co-governance of territories along with the protective spirit beings and other-than-human beings of the lush beauty and diversity of life. In the Living Forest, nonhuman others have a priority role in cocreatively shaping conviviality in nonhuman-centric terms. This communality among other-than-humans, wherein humans are co-responsible, underpins Indigenous living practice according to a logic and spirituality not understood by the West. This is a proposal of conviviality based on self-determination and self-governance of territories that has its own institutionality and is against extractive projects and their state sponsors (Angun [Jose] Gualinga, Kichwa People of Sarayaku).

Angun Gualinga explains that the Kichwa People of Sarayaku have sought to transfer this into a political proposal of the sovereign self-governance of the Living Forest. In this proposal, humans are only a part of broader Plans of Life. This is very different from the (human-centric) state plans of development on the one hand and protected areas on the other. Indigenous living territories do not spatially separate other-than-human from human thriving but were originally instructed to co-nurture each other. Hence, the belief in autonomous and full governance is rooted in the sovereignty of the Living Forest herself. In transferring this into the dominant logic of the state, Sarayaku has been involved in judicial battles. For example, in relation to oil extraction, the people of Sarayaku have won a case (over 10 years ago) in favor of their community territories and against extraction. Yet the measures won in the courts (including the Interamerican Court) have not been honored, and state regulation favoring Sarayaku has not been implemented. There has been no effective free, prior and informed consent, and customary law, free self-determination and Indigenous law have not been respected. Communities like the Kichwa People of Sarayaku underline consent based on their Indigenous norms and customary governance. In Indigenous law, consent demands, of course, the consent of the Indigenous (human) community, but also that of the other-than-human community of beings and spirits. But here is where the macro-logic of the state (anthropocentric, patriarchal, Eurocentric, rationalistic, positivistic, and objectifying) overlooks the situated personhood of Indigenous living territories and communities as embodied by Indigenous people's cosmovision and practice. Indigenous peoples like the Amazonian peoples have real solutions rooted in millenarian living experience, outside and beyond the dominant logic. Western civilization has considered Indigenous ways only as traditions, (even as worthless traditions) treating them as museum artifacts of a dying past, subsuming them under its own logic and law, or treating them as "local" supplements to its universal science. But Indigenous lifeways stand on their own as embodied philosophical and spiritual practices, understood as cosmovisions for originary (Indigenous) peoples, now in political transformation (Angun [Jose] Gualinga, Kichwa People of Sarayaku).

Angun Gualinga further exemplifies that a different lifeway and knowledge stems from acknowledging and relating to Nature as a conscious living being. Indigenous cosmovisions, such as those of Amazonian peoples, call us to live in harmonious connection by acknowledging the complex thought and agency of other living beings and their languages. These Indigenous cosmovisions are not "traditions" but embodied presents with deep continuing histories and transformative proposals for viable futures that confront and threaten the racist, capitalist, and consumerist world. The problems that Indigenous peoples and communities confront and the solutions they embody carry thousands of years of experience that now become an *other* politics rooted in broader living communities, like the Kawsak Sacha. These constitute horizons of life as the Sumak Kawsay, well beyond capitalist "development" and consumerism. The Sumak Kawsay requires a communal effort to strengthen convivial harmony on the basis of acknowledging and listening to the complex thought of nonhuman and human beings, whether visible or invisible to the senses. TIAM is a vision of transformation oriented toward the common house that re-situates the local in a planetary context. The Kichwa People of Sarayaku have been working to strengthen Indigenous institutionality and self-government in accordance with Indigenous cosmovision, normativity, and community council decision-making (Angun [Jose] Gualinga, Kichwa People of Sarayaku).

Çaca Yvaire of the NEFOC articulates that the four distinct processes do not lose their situated meanings as they are interwoven in a common fabric of alternatives. For example, the practice of the Living Forest (articulated by Sarayaku) and other communities/territories of life weaves our processes in relations with each other in their acknowledgment of the sovereignty and self-governance of territorial communities of life. Certainly, we see similarities in our opposition to a violent global logic and all processes acknowledge territories as sovereign, living, and agential biocultural ecosystems. NEFOC has been focused on liberating lands on their own terms, as well as liberating subjugated peoples by facilitating the rematriation of lands to Indigenous and people of color. NEFOC has sought to facilitate the decolonization process in co-envisioning with such communities the complex task of navigating the settler colonial state's legal regime. At times, simply trying to exist and subsist as communities and organizations subject to an order that forces upon us the need to be legally and financially legible in settler colonial-capitalist terms is challenging.

This notwithstanding, one particular aim has been to nurture relations with and among communities so that their modes of land access proceed to overcome the colonial structure and settler property regimes, and that these processes of land liberation and rematriation can weave a network of lands that enacts communal conservation based on the principle of land as sovereign. In this context, NEFOC processes recognize themselves in the reflection of the work done by the companions of Sarayaku and their recognition of the agency and sovereignty of the Kawsak Sacha. NEFOC works to underline the principle of the sovereignty of ecosystems, which is the basis of knowledge sharing with BIPOC communities and farmers seeking land rematriation, reparation, and access. The notion that ecosystems are sovereign, living, spiritually active, and cocreative agents that govern themselves, co-organize, and engender themselves and take priority in shaping how we ought to live collectively is key to an other politics, an other conviviality that does not reinscribe notions of mastery, dominance, and property over others and over land, and instead cocreates worlds where creativity is shared among diverse beings. We need to emphasize the centrality of other-than-human agency and knowledge, which precede and enable human life and without which human life cannot be possible.

Certainly, as the companions from Sarayaku, ASPROCIG, and NEFOC underline, our visions are not of commodification but of playing our responsible role in the wider community of cocreation to nurture and nourish the beauty of our big home. We are emphatically not "stewards" of the nonhuman but descendants, relations, and students of other-than-human spirits and beings. Humans are in many ways learning our responsibilities from those teachers, other-than-humans, and cosmic forces who were here before us and whose collective relations and actions make us possible in the first place. To them, we owe responsibility and reciprocity, and without their goodwill we cannot be. Here it is key to center a non-anthropocentric Earth-center jurisprudence based on multispecies relationality, agency, and interagency. The transition among actually existing and aspiring BIPOC farmers toward communal land rematriation outside of the setter-colonial state logic proceeds from the reestablishment of a convivial relation with other species that recognizes their personhood and knowledge. These are delicate processes because in reconstituting and/or creating BIPOC farmer communities we confront the imperial present where colonial and capitalist forces intervene in territories, structure orders, and rupture relations in ways that limit collaboration (Çaca Yvaire, NEFOC).

For NEFOC, Çaca Yvaire notes, this is patently clear since its work seeks to unsettle colonial, capitalist, and heteropatriarchal regimes in the very core of empire, the United States, whose universalizing and globalizing logic enables a power that is translated into impacts replicated across all territories. Like other processes in this interweaving exercise, Indigenous and Afro-descendant communities have been divided and counterpoised by settler imperialism, in this case the US empire. On the one hand, the history of slavery among Black peoples and their struggle for liberation, justice, and reparations has often been divided from the history of genocide and displacement of Indigenous peoples and their struggle for sovereignty, decolonization, and rematriation. We confront a combined history of colonial genocide, the imposition of a settler colonial state property regime, enslavement, systemic racism, and heteropatriarchal domination. We must confront the fact that reparations for Black slavery must happen in the very same territorial space as Indigenous land rematriation. Thus, returning to the roots, in land and food sovereignty, means not restoring a pristine romanticized past but laboring together in celebration of shared struggles, dreams, and visions to create new soils. This must include contributions from others, as there are also arrivants here from the peripheries of empire, dispossessed (often Indigenous and Black) diasporic communities forced from their territories by the tentacles of empire. Together, Black and Indigenous communities must work, particularly centering women and non-binary leadership against heteropatriarchal property regimes, to build new soils, interwoven by branches from different roots which can give life to rebirthed and liberated territories. Yet, what this dominant system has induced all along is a deterritorialized hopelessness where peoples are uprooted from a convivial relation to territorial communities of life (e.g., from slavery and colonialism to contemporary settler colonial, racist, capitalist, and imperialist dispossessions). Therefore, processes like NEFOC center the co-exploration and coordination of struggles for reparation and rematriation. This healing against lived racism is built on healing from the separation and alienation from nature. It requires that we recognize the cocreative agency of sovereign ecosystems and nonhuman communities and proceed to weave our struggles into an emancipatory fabric rooted in a deep respect for differentiated yet profoundly interrelated histories upon the local and situated agency of communities of life. Here both ecosystems and peoples must be recognized in their distinctive personhood, agency, and particularity.

Yet, such a planetary sensibility that is impressed upon each of our processes by the sovereignty and agency of land, Earth, and the other-than-human must be differentiated from a global logic. Weaving among distinct communities of life while retaining their identity distinct from the global logic, it is key to understanding our shared planetary comprehension which interlaces locality in a collective movement. Understanding each other as part of an interlaced planetary web also means proceeding from a strong recognition of the contextual differences and locations in a deeply uneven planetary landscape of power and inequity. NEFOC, for instance, is situated in the core of late-modern empire, and from here stems the entrenchment of processes of recursive accumulation harming others. Hence, disrupting/uprotting power here is key to helping unravel the imperial tentacles of power intervening in other territories. Being at the core of the empire, we must unravel the logic of setter dispossession underpinning imperial expansion across territories. Thus we can redirect lands and resources as we try to redirect power (Çaca Yvaire, NEFOC).

Weaving the Territorial Fabric of Life
beyond a Civilization in Crisis

This chapter results from a cocreative process of weaving a tapestry of emancipatory communal lifeways through a collective intercambio de saberes. This intercambio gathers the voices of four key processes from across Abya Yala and Turtle Island: the Kichwa People of Sarayaku's Kawsak Sacha/Living Forest; The Association of Indigenous and Afro-descendant Fishermen and Peasants for the Community Development of the Bajo Sinú (Colombia); EDUCA, A.C., Services for an Alternative Education's campaign for Communal Alternatives in the Defense of Territory (Oaxaca, Mexico); and NEFOC, Northeast Farmers of Color Land Trust (Turtle Island/United States). Each of these diverse, locally situated voices is grounded in autonomous communities nurturing rich biocultural territories. These constitute actual horizons for sustaining, diversifying and propagating multispecies communities of life. The processes, which are exemplary nodes of a much more extensive fabric of communal defense and regeneration, do not embody "alternatives." Rather, they are persistent lifeways that precede, resist, survive, and will exceed a modern/ colonial civilization in decay. They reconstitute communal multispecies territories of self-determination rooted in cosmovisions and collective practices that articulate Indigenous, Afro-descendant, Afro-Indigenous, peasant, fisherfolk, and forest peoples. They are constituent members of territories that embody visions of land, food, knowledge, and identity sovereignty and self-determination, and whose collective practices unfold into emancipatory horizons. These horizons are incommensurable and irreducible to a global or universal logic, but enact a pluriversal fabric of complementary lifeways. These lifeways return and recreate themselves to materialize actually existing worlds beyond an anthropocentric, patriarchal, colonial, capitalist, state-centric, and developmentalist model imposed upon a violent logic of global extraction. The cocreative agency and spiritual-material autonomy of the communal territories envisioned and pursued by these four processes (and the other-than-human, human, and land/waterscape persons that relationally regenerate them) are rooted in the recognition and respect of the sovereignty, complex intelligence, and living agency of biocultural territories and multispecies communities. These enact, precede, and exceed the regeneration of human governance which must be practiced as just one second-order part of non-anthropocentric communal action, and through co-learning and co-responsibility with the extended fabric of other-than-human relations that make the cycles of regeneration possible. The celebratory and cyclical reenactment of life in its plentiful richness, diversity, and vitality must be grounded in (1) the sovereignty of the land and Mother Earth, (2) the deep spiritual acknowledgment of other-than-human personhood, cocreativity, intelligence and agency, (3) the self-determination of communities, and (4) the exaltation of a *vivir sabroso* (a delicious living) whose beauty results only from the rekindling of communal energy, *Sumak Qamaña/Suma Kawsay/Lekil Kuxlejal*—supreme energy-spirit-vitality cyclically regenerated by reciprocal conviviality among all community members.

9

Lifting up the Liberation Praxis of Africana Indigenous Peoples, Peasants, and Urban Slum Survivors

Samuel Leguizamon Grant

Introduction

This chapter offers an introductory examination of three social formations actively organizing against ecological imperialism: Indigenous peoples, peasants, and urban slum dwellers. These social formations will be looked at through the experiences of Africana people (people of African descent on the continent and across the diaspora) and through the lens of Africana critical theory (Rabaka, 2009, 2010a, 2010b, 2014, 2020). Africana critical theory proposes there are, inevitably, two linked processes of liberation praxis: revolutionary re-Africanization (resurgence of anti-imperialist culture) and revolutionary decolonization (global linkages among anti-imperialist struggles). We will look at ways in which these two processes manifest in the liberation work of Africana Indigenous peoples, peasants, and slum survivors.[1]

Africana Indigenous peoples, peasants, and slum survivors around the world are building on traditions of principled struggle, seeking ways to build power to do more than survive capitalism-imperialism. They organize to resist and shape alternatives that foster cultural and intercultural well-being and mutual liberation. By defending Indigenous lands and livelihoods, we protect the remaining biodiversity on the planet and learn how to organize the collective resurgence of local ecological knowledges embodied in local ecological democracies. By supporting the food sovereignty objectives of both Indigenous peoples and peasants, we grow a replacement for the industrial food system

---

[1] Social Movements Institute on Living Ecologies (SMILE) and Rainbow Research.

controlled by capitalist monopolies and learn to feed ourselves within local living cultures. This ensures an abundance of healthy local food is available, even in times of crisis. By supporting the campaigns of slum survivors, we radically increase safety and community and generate novel, just transition initiatives from the bottom up, transforming urban slums into ecologically democratic oases.

However, these movements are not linked either at local or regional scales, except through discourse at World Social Forums, which only a scant few of the world's poor can ever attend. As we consider a transformed future in the twenty-first century in which the land is healthy, the people are healthy, and we have cocreated and coevolved infrastructures for cultural and common well-being, I argue that the leadership of people from Indigenous struggles, peasant struggles, and slum survivor struggles will be critical to ensure the best possible outcomes for all of us. Africana Indigenous peoples, peasants, and urban slum survivors are actively resisting and proposing alternatives to capitalism-imperialism.

As Walter Rodney put it, people "pan" up (as in Pan-African) as "an exercise in self-definition by a people, aimed at establishing a broader redefinition of themselves than that which has been permitted by those in power" (Rodney, 1975, p. 18). The extent to which social movements of Africana Indigenous peoples, peasants, and slum survivors can link with each other and with intercultural social movements at world scale is key to significant positive transformations in the twenty-first century.

Capitalism-imperialism continues to gobble up land, water, and minerals under the earth, and destroy the life scapes of Indigenous peoples, peasants, and slum survivors. It imprisons us all in various necropolitan realities (Mbembe, 2003, 2019), in which lives and dignity are disordered because they are in the way of capital accumulation. Fanon (1967) characterized this function of colonialism as throwing us out of our self-determining realities and into the "zone of nonbeing." Black bodies and lands are situated as sacrifice zones for capitalism-imperialism, and it is currently going from bad to worse.

Samir Amin asserts that capitalism is in an L crisis—meaning permanent decline. "Decline is a very dangerous time. Capitalism will not wait quietly for its death. It will behave more and more savagely in order to maintain its position, to maintain the imperialist supremacy of the centers" (Amin, 2019b, p. 13).[2] We see this play out in the massive land grabs across the African continent

---

[2] Amin makes the additional point that capitalism is fading into fascism at world scale. This means that we will increasingly see manifestations of fascism-imperialism replacing capitalism-imperialism.

and diaspora, and the rising tide of dispossession by gentrification in the Global North, adding to the utilization of the prison industrial complex, denial of voting rights, and refusal to collectively embody and work from a critical, intercultural understanding of our linked fates as human beings and as earthlings.

"Imperialism is the system by which a dominant power is able to control the trade, investment, labor, and natural resources of other peoples" (Frame, 2022, p. 508). Ecological imperialism is the assertion of power and control by some social formations (capitalists and imperialists) over the land and lives of people in other places all across the world. It includes the social reproduction of this power in ways that both facilitate economically and ecologically unequal exchange, and impose greater negative economic and ecological impacts across the African continent and diaspora (Frame, 2022, pp. 508–509). Since the mid-fifteenth century, slave trade began on the continent of Africa; ecological imperialism has facilitated the development of the world capitalist system on the backs and bodies of Blacks, using us as the fuel for accumulation alongside carbon. As capitalism and imperialism combine (capitalism-imperialism) to facilitate ecologically and economically oppressive exchange, people in places negatively impacted organize forms of resistance and work to sustain and/or generate alternatives (Frame, 2022, p. 509).

The questions explored here are: How are Africana Indigenous peoples, peasants, and urban slum survivors adapting to and working to transform the impositions of ecological imperialism, and what can we learn from their struggles to advance common liberation? First, however, we need to answer a preliminary question: Why look at these three social formations in the context of countering ecological imperialism?

Indigenous peoples, peasants, and urban slum survivors are all engaged in land-based struggles, and they are all constantly among the most impacted by the processes of accumulation by dispossession, which is the basic expression of ecological imperialism. Colonial imposition, which ushers in the work of capitalism-imperialism, forcibly removes Indigenous peoples from their lands,[3] turns them into peasants, who are then, all too often, turned into urban slum dwellers[4]. These three social formations counter these impositions through (1) Indigenous land defense by various means, (2) land occupations and organizing

---

[3] The violent transition forced by capitalism-imperialism from land-based Indigenous self-determination is grounded in genocide—far more have been killed than have been forcibly removed to rural or urban locations.

[4] Mike Davis's (2006) *Planet of Slums* tells the story of urbanization without opportunity as among the most critical challenges of this century.

Table 9.1 Amin's Five Monopolies and Peasant, Indigenous, and Urban Poor Resistances Countering Them

Amin's Five Monopolies	Peasant Struggles	Indigenous Struggles	Urban Poor Struggles
Natural Resource Control/ Direction	Land Occupations Farmer-to-Farmer linkages: developing and sharing agroecological knowledge and partnering to protect against further assaults of land grabbing from industrial agricultural interests	Fighting to protect and restore land to the commons	Engaging in innovative strategies to secure water, obtain electricity from the grid, and challenge the grid to recognize informal settlements
Financial Control/ Direction	Countering financial constraints through community-supported agriculture and local farmers markets. However, these efforts are significantly entrapped by a lack of autonomous financing at all scales	Maintaining the decommodified nature of Indigenous livelihood is a significant source of financial autonomy and cultural protection	Engaging in creative modes of local economic organization that allow people to meet their basic needs. However, these efforts are always subject to being undermined by state forces and capital

Information Control/ Direction	By making direct linkages to communities through peasant associations and challenging the state to respond favorably to peasant and communal demands, peasant struggles reverse information control/direction	Developing Indigenous education systems and fostering self-determined forms of "development" based on ethics of cultural, resource, and land-based sovereignty	Poor people's movements are linking globally, building a common knowledge base, and fostering emerged forms of shared praxis for collective liberation
Technology Control/ Direction	The internet age has allowed peasants to learn with and from each other in online formats	Braiding modern appropriate technology with Indigenous technologies	The internet age has allowed the urban poor to meet synchronously and learn from each other asynchronously
Military Control/Direction	Here is the space where the violence of ecological imperialism is devastating, and we lack sufficient answers to protect existing gains and keep escalating the struggle for common liberation until all are both free and well		

for food sovereignty by peasants, and (3) organizing among slum dwellers for alternative ways to provide basic needs and support each other with emergent strategies of community governance.

We can metaphorically, as well as pragmatically, support and link the gifts of these social movements for common liberation. From Indigenous peoples and peasants, we can work for the restoration of traditional ecological knowledges, diverse seed varieties, and cultural practices that honor all of us as relatives on the land. From slum survivors we can learn the grit and tenacity it takes to carve out space in the literal belly of the beast—organizing for liberation in massive urban conglomerations where there is a daily struggle for sanitation, water, food, and economic opportunity, as well as constant assaults on dignity and well-being.

To contextualize the liberation praxis of Africana Indigenous peoples, peasants, and urban slum dwellers, I will examine cases by looking at how they work to counter the impositions of the five monopolies of capitalism-imperialism as theorized by Samir Amin (2011, 2014, 2019a, 2019b). Table 9.1 juxtaposes Amin's five monopolies against what the movements of Indigenous peoples, peasants, and slum survivors offer as forms of resistance and alternatives. This picture is incomplete and will be further developed. The key point is that eradicating the stranglehold of monopoly power in the present and future is a key foundation of struggle across social movements that recognize that they don't get to have a future worth having under the constraints of such powerful and manipulative control and destructive capacity.

Afro-Indigenous Organizing

In 2007, the United Nations authorized the historic Declaration on the Rights of Indigenous Peoples—UNDRIP (UN Department of Economic and Social Affairs, 2019). While the worldwide Indigenous peoples' movement was slower to catch fire on the African continent, it is growing in significance. Indigenous peoples in Africa include hunter-gatherers, numbering around 400,000, and pastoralists, which number over 268 million, according to USAID (2021: 4). The United Nations estimates a total of 50 million Indigenous people in Africa (UN Permanent Forum on Indigenous Issues, 2013). Since the 2007 declaration, a few positive outcomes have emerged within the context of ecological imperialism's nexus of institutions, including the UN (State of the World's Indigenous People V4, 2019: 11). These include the following:

- The 2006 Constitution of the Democratic Republic of the Congo recognized minority cultural group rights (Mochire et al, 2022).
- In 2010, the Central African Republic ratified the International Labor Organization (ILO) Convention 169 (Rainforest Foundation, 2009)—the precursor to the UNDRIP which recognized the human rights of Indigenous peoples. It became the first African nation to do so (Permanent Forum on Indigenous Issues, 2013).
- In 2011, the Democratic Republic of the Congo was the first African nation to formally recognize the rights of Indigenous peoples in law.
- Niger and Chad developed pastoral codes in 2012.
- Namibia, in 2014, began thinking through how to honor the UNDRIP with Indigenous peoples in the region.
- Through the work of the Alliance for Food Sovereignty in Africa, many case studies are being generated that describe how Indigenous peoples and peasants are saving seeds, protecting, and restoring culture (Gebremikael, 2019; Pinto, 2019);
- Indigenous knowledges are lifted up, restored, and connected with conventional science to support self-determination and respond to the climate crisis (Gebremikael, 2019; IPACC, 2021).

In addition, several other Indigenous organizations, struggles, and processes have spread across the continent. For example, the Indigenous Peoples of Africa Coordinating Committee (IPACC) Council is a network of 135 Indigenous peoples in 21 African countries. Their work documents displacement and land grabbing, and also works constructively to increase the integrity and commitment to the practice of free, prior, and informed consent, and to train Indigenous peoples in conflict mediation (IPACC, 2023). Through the IPACC, action campaigns on several fronts work to build local power of Indigenous groups, link them together in common cause, and compel both national and global governance to respect Indigenous lands and lives more fully.

Another example is the African Commission on Human Rights, an intermediary charged with addressing the objectives and concerns of Indigenous peoples on the continent (UN DESA, 2019: 18). Indigenous peoples organized and challenged the African Commission on Human Rights to recognize their interests beginning in 1999. This resulted in the beginning of internal work within the commission in 2000 (Sena, 2015: 1). In 1989, Maasai elder Moringe Ole Parkipuny became the first Indigenous person in Africa to address the United Nations (Sena, 2015: 2). Even though it has limited funding and no real enforcement power, the African Commission on Human Rights does have influence through the use of international pressure to bear on the decisions and processes of nations on the continent. In two cases, the African Commission has had an impact—the Endorois and Ogiek struggles to have their land base restored in Kenya (UN DESA, 2019: 33).

The Indigenous people's movement continues to gather energy and power in Africa but is hampered by limited resources, repression, and internal political strife (Sena, 2015: 2). As a result, their struggles are marked by both successes and failures. For example, the San in South Africa were "awarded" land rights to limited land in 1999, yet three years later, this land was put under a form of administrative control that prevented the San from practicing livelihoods on their terms (Jansen, 2020). However, in 2009, the Khoi and San were recognized as the traditional ecological knowledge holders of "Rooibos" (Natural Justice, 2019). An agreement was finally concluded in 2019 that offers them a percentage of profits from the commercialization of the crop (Jansen, 2020). President Ramaphosa also signed the Traditional and Khoi-San Leadership Act in 2019.

Nonetheless, Indigenous peoples in Africa continue to face significant threats, such as the Forest Peoples of Central Africa who now number 300,000 or less and live under significant cultural and ecological threat by predatory governments and multinationals (Carino, 2009: 34). Similarly, there remain 87,000 or less Batwa people, only 7,000 of whom can sustain regular access to the forest to support their livelihood (Carino, 2009: 34). The climate crisis is also escalating threats to Indigenous livelihoods. Not only do they face the environmental devastation of climate change, they also face new threats as international institutions and governments work with national governments in Africa to develop carbon markets as a key way of addressing climate change. Studies have shown how such actions exacerbate accumulation by dispossession and the destruction of Indigenous livelihoods (No REDD in Africa Network, 2015). Consequently, Indigenous peoples are standing up to initiatives, such as Reducing Emissions from Deforestation and Forest Degradation (REDD), criticizing them as "false solutions" which socially reproduce imperialism (Goldtooth, 2019).

The above provides some context for the struggles of Indigenous peoples in Africa, but it is important to look at some examples of their organizing. The Afar in Ethiopia and Eritrea (numbering nearly two million people) have sustained a distinct culture and the institutions that support their own form of ecological democracy (Gebremikael, 2019). Afar traditional ecological knowledge (TEK) is proving critical in light of climate change. Now traditional culture keepers are working with universities to combine TEK with "conventional" science locally and globally (Gebremikael, 2019).

The IPACC is doing its best to facilitate the integration of traditional ecological knowledge into climate science and negotiations about how the world's nations respond to the climate crisis. They identify many practices of Indigenous ecological knowledges relevant for all of us as Indigenous peoples respond to climate change. These include observing changes in animal and plant behavior, responding to climate threats with local ecological governance, varying their diets and food intake in response to changing conditions, proposing alternative ways of assessing hydro-ecological conditions, effectively using bioindicators to forecast weather patterns and impacts, and much more (Crawhall, 2016: 8). Further, IPACC has made recommendations to African governments on how they should work with Indigenous peoples. Among these recommendations are (1) supporting Indigenous community control of the lands in which they live, (2) conserving wild resources and ecosystems under a comanagement paradigm with Indigenous peoples, and (3) following Indigenous leadership in terms of how Indigenous ecological knowledge and other knowledges are merged to respond to climate change (Crawhall, 2016: 15).

What we can minimally discern from looking at this small sample of Afro-Indigenous struggles as they work to counter monopoly power is that the insight, capacity, and linkages across Afro-Indigenous movements are growing. The forces are not capable of challenging capitalism-imperialism on their own. They need slum survivors and the global peasant movement to join them as full partners in solidarity ecologies that organize around mutual benefit for common liberation.

African Peasant Organizing

Peasant movements around the world exemplify resistances against ecological imperialism through struggles for food sovereignty and land reclamation. La Via Campesina (1996, 2007a, 2007b, 2008, 2013), one of the foremost internationally linked peasant movements in the world, articulates that bottom-up agroecology can feed the world and replace the industrial agricultural system, which generates hunger, obesity and food-borne diseases in both the Global South and obesity and the Global North (FAO et al., 2023). Miguel Altieri (2012: 2), a leading agroecology expert, asserts that a "more radical transformation of agriculture is needed, one guided by the notion that ecological change in agriculture cannot be promoted without comparable changes in the social, political, cultural, and economic arenas that determine agriculture." Altieri (2009: 7) continues: "small, diversified farms can produce 2 to 10 times more per unit area than do larger corporate farms." Unlike strictly capitalist small-scale farmers, peasants provide for themselves and primarily grow food for their local communities. The United Nations, after much delay, adopted the United Nations Declaration on the Rights of Peasants in 2018 (UNDROP). The declaration distinguishes peasants from capitalist producers based on their utilization of non-monetized social labor as part of production, a focus on community production and the sustaining of a strong attachment to the land (Article 1 of UNDROP).

Peasants around the world are engaged in land back movements of their own. For example, Sadomba (2013: 97) reports on the long struggle of Zimbabwe's war veterans to take back land stolen by Rhodesia. Their land was not restored until the more than two decades of struggle against the government waged by the national liberation forces of Zimbabwe, which secured independence in 1980. War veterans "won" compensation and made possible the return of land from white colonizers to African peasants. No other African country has dared to try a similar strategy or had a group like the war veterans with the organized capacity to force the issue.

The National Council for Rural Concertation and Cooperation (CNCR) in Senegal emerged in 1993 and has worked nationally to organize the voice and power of peasants and defend their rights and interests as producers (McKeon et al., 2004: 8). The CNCR emerged via the springboard of a national peasant association, FONGS (Federation des ONG senegalaises), which formed in 1978 (McKeon et al., 2004: 11). The basic building block of the CNCR strategy was to restore a sense of peasant responsibility for their own capacities in relation to each other, their communities, the state, and the world (McKeon et al., 2004: 15). It is in relation to the base communities that CNCR is most challenged as "very few federations have structured membership and communication chains" (McKeon et al., 2004: 20). And, of course, this peasant association, like most others, finds it difficult to raise and sustain funds to do all it wishes to do on behalf of peasants and their communities (McKeon et al., 2004: 25).

The Rural Women's Assembly made a declaration in 2014 asserting that "rural women in Africa are the main producers of food, yet their contributions remain invisible. They are the most marginalized in terms of access to land and secure tenure, natural resources, and political rights. Patriarchal relationships continue to prevail making rural women vulnerable and subject to violence" (Rural Women's Assembly, 2015a: 3). In 2015, the Rural Women's Assembly made a clear demand: "one woman—one hectare of land!" (Rural Women's Assembly, 2015b: 1). In addition, they demanded: (1) the provision of commonage to poor people, rather than leasing it out to commercial farmers; (2) secure and independent tenure and land access for farm workers/dwellers, including women and seasonal workers; (3) women's independent tenure security and land ownership; (4) housing contracts in the name of women farm workers; (5) an immediate end to farm worker evictions; (6) prosecutions of farmers guilty of constructive and illegal evictions; (7) provision of emergency accommodations when farm workers are legally evicted; (8) an end to ineffective farm worker equity schemes, which make false promises to farmers and leave them vulnerable; (9) dignified housing with water, sanitation, electricity, and other basic services; (10) enforcement of minimum wages for farm workers; (11) rights of small-scale farmers to land and water; (12) environmentally sustainable agricultural support to small-scale farmers; (13) an end to policies and practices which discriminate against women in traditional authority areas; (14) preferential access and rights for artisanal and subsistence fishers; and (15) the right to use forestry plantations for income generation for forestry communities (Rural Women's Assembly, 2015b: 2).

As Moyo (2013) points out, these struggles are (like all resistances) marked by both successes and major limitations. Moyo identifies several important contradictions, including the fact that movements are expected to (1) work through liberal civil society reliant on private donors; (2) fight for multiparty democracy even though all "acceptable" parties have given in to neoliberal logics; and (3) to respect the "rule of law," which privileges private property rights over the commons and traditional tenure.

What can we glean from the small number of cases from the African agroecology peasant struggle presented above? The peasant struggle on the continent is gaining in wisdom and power. The farmer-to-farmer organizing and linking across regions is building an increasingly clear anti-imperialist strategy. However, it currently lacks the ability to displace monopolistic power. To get there, it needs to organize for the formation of linked struggles across social movements with Indigenous peoples, slum survivors, and all the other folks who recognize that there can be no good common future with imperialism.

African Urban Poor Organizing

Urban slum survivors have been forced through processes that have removed them from their land and land-based culture, deemed them unworthy of available work, and forced them to find a place to exist without support from either state or capital. They exist in a *Black ecology* (Hare, 1970) of living in slums that possess the "attritional lethality" of slow violence (Nixon, 2011). Pain (2018) theorizes that the constant experiences and threats of future dispossessions keep the urban poor in a state of "chronic urban trauma." This is also analyzed by psychiatrist Dr. Mindy Fullilove as "root shock." The violent interplay of Blackness and geography forms the core root of capitalism-imperialism (McKittrick, 2011). Twenty percent of the world's urban slum population (1 billion people) live in Sub-Saharan Africa (Ganz, 2020). At nearly 70 percent of the urban population, they represent the greatest concentration of urban poor in the world and face the highest deficit in terms of affordable housing supply compared to demand (Bah et al., 2018).

South Africa is a key site of slum dweller organizing. Ayanda Ngila, leader of the Ekhenana Commune—one among many communes of the shack dwellers movement in South Africa—was assassinated in March 2022. Ekhenana is a democratically run housing settlement, commune, and farm. "Ayanda was assassinated wearing a Steve Biko t-shirt with the writing 'it is better to die for an idea that is going to live than to live for an idea that is going to die'" (Dalilah, 2022b: 2).

Abahlali baseMjondolo formed in 2005 to challenge evictions in the Kennedy Road Settlement (Durban, South Africa) and has grown to include more than 100,000 members (Dalilah, 2022b: 2). Recognizing that the revolution will not be funded, Dalilah (2022b: 3) asserts that "it is the boldest, most daring movements with the capacity to make the most radical, impactful change are those which are perceived as the highest risk to fund." The movement refuses to be halted because people refuse to die or lay down for a false peace. Abahlali baseMjondolo, working with the Amadiba Crisis Committee, the South Durban Community Environmental Alliance, and others, has successfully halted the quest of Shell Oil to hunt for and extract oil from Kwa-Zulu Natal's wild coast (Dalilah, 2022a: 2). They won by a mix of strategic litigation, taking it to the streets, boycotts, petitions, and challenging Shell on its failure to practice appropriate corporate responsibility (Dalilah, 2022a: 2–3).

The South African Thembelihle Crisis Committee (TCC), after exhausting local remedies to ensure basic needs of slum settlements, took their case to the regional authority and staged a three-week road blockage and strike. This ultimately resulted in formal recognition of the settlement and set the stage for regular access to water, sanitation, electricity, and housing (Sinwell, 2017: 17). Similarly, the 2012 uprising, following the massacre of thirty-four workers at Marikana in South Africa, demonstrates the significance of evolving forms of struggle. In this case, linking the autonomous organizing of workers with the autonomous organizing of base communities working in active solidarity with each other (Ngwane, 2017: 47).

If the state and capital fail to honor the dignity of all lives and ensure basic needs are met, people will persistently struggle for rights and freedom from the violent impositions of imperialism. The constant movement activity (at micro and meso-scales) of Abahlali base Mjondolo and the numerous land occupations asserting the autonomous right to exist, to survive, and to find ways to thrive offers an important example of people's struggle that will continue throughout the twenty-first century (Ngwane, 2021). In more and more cases, people are no longer waiting for handouts or handups—we are taking our destiny in our hands and countering the ambition of imperialism to foreclose all alternatives.

Transforming Potentials?

Amílcar Cabral theorized that imperialism "cannot be sustained except by the permanent and organized repression of the cultural life of the people in question. It can only firmly entrench itself if it physically destroys a significant part of the dominated people" (Cabral, 1974: 12). He further theorized that it is only in culture that the seeds that foster liberation are planted and nourished to grow. The social movements of Africana Indigenous peoples, peasants, and slum survivors are all planting seeds of liberation in their local struggles and learning to connect with each other at a global level. However, there is not yet sufficient cocreative liberation praxis that joins the work of Indigenous peoples, peasants, and urban slum survivors. This is the next necessary stage of struggle.

Globally, Indigenous people constitute 5 percent of the world's population and are largely unauthorized custodians of 80 percent of the world's remaining biodiversity (Secretariat of the United Nations Permanent Forum on Indigenous Issues, 2013: 1). Given the failure of ecological imperialism to work within biospheric limits (Meadows et al., 1983), one of the most important things we can do in the world now is to fully honor the world's Indigenous people as our collective authorities regarding the protection of nature and ecosystems on which all human and other-than-human life depends.

Peasants compose close to 1.2 billion people[5] in the world (Hubert, 2019). Together with their families, peasants compose one third of the global population (Hubert, 2019). Support for food sovereignty, following the leadership of rural and urban peasants, can increase the availability of local, healthy, and affordable food everywhere in the world and reduce the massive carbon footprint of industrial agriculture.

Urban slum survivors compose more than 1 billion[6] people in the world. As they work to challenge local and state governments to commit to development without displacement and to support the leadership of slum survivors in urban planning for just transitions in response to the climate crisis, they help transform core urban conditions from accumulation by dispossession toward urban ecological democracy.

Samir Amin argued that the objective of national movements in the Global South must be to delink from the expansionary project of ecological imperialism (Amin, 1990a, 1990b). I argue here that what makes more sense now is a commons-based struggle among organizing populations not-aligned with the state or capital but aligned with each other around common objectives. African Indigenous peoples, peasants, and urban poor around the world are building on traditions of principled struggle, seeking ways to ensure they amass the power to pass on a better future to coming generations of Africans. As Hare (1970: 8) put it, "the real solution to the environmental crisis is the decolonization of the Black race." Rabaka (2009) would add that in order to decolonize, Black people on the continent and across the diaspora need to embody a resurgence of decolonizing, anti-imperialist cultures and link cultural and intercultural struggles for mutual liberation. McKittrick (2011: 960) further asserts that better praxis will emerge when we "re-imagine geographies of dispossession and racial violence as sites through which 'co-operative human efforts' can take place and have a place."

In the end, it comes down to embodying and collectively operationalizing a set of anti-imperialist policies and practices that displace and dispossess capitalism-imperialism of power over our lives and lands (Amin, 2019b 25). Let us mutually move all that is necessary for this to happen into being, into all relations, and into all actions until we can celebratorily embrace a world of common conviviality.

---

[5] https://www.cetim.ch/rights-of-peasants/#:~:text=At%20the%20global%20level%2C%20there
    ,represent%20one%20third%20of%20humanity.
[6] https://unstats.un.org/sdgs/report/2019/Goal-11/

References

Altieri, M. 2009. *Agroecology, Small Farms, and Food Sovereignty*. Monthly Review. https://monthlyreview.org/2009/07/01/agroecology-small-farms-and-food-sovereignty/#:~:text=Rural%20Social%20Movements%2C%20Agroecology%2C%20and%20Food%20Sovereignty&text=A%20more%20radical%20transformation%20of,arenas%20that%20help%20determine%20agriculture

Altieri, M. 2012. "The Scaling up of Agroecology: Spreading the Hope for Food Sovereignty and Resilience." *SOCLA*. https://archive.foodfirst.org/wp-content/uploads/2014/06/JA11-The-Scaling-Up-of-Agroecology-Altieri.pdf

Amin, S. 1990a. *Maldevelopment: Anatomy of a Global Failure*. London: Zed Books.

Amin, S. 1990b. *Delinking: Towards a Polycentric World*. London: Zed Books.

Amin, S. 2011. *A Global History: A View from the South*. Nairobi, Kenya: Pambazuka Press.

Amin, S. 2014. *Capitalism in the Age of Globalization: The Management of Contemporary Society*. Zed Books.

Amin, S. 2019a. "The Long Revolution in the Global South." *Monthly Review Press*.

Amin, S. 2019b. "Globalization and its Alternative: An Interview with Samir Amin (Part 1–3)." *Tricontinental Institute for Social Research*. http://www.socialisteconomist.com/2019/01/globalization-and-its-alternative.html

Bah, Eh. M., I. Faye, and Z. F. Geh. 2018. "The Housing Sector in Africa: Setting the Scene." In *Housing Market Dynamics in Africa*. London: Palgrave Macmillan. https://doi.org/10.1057/978-1-137-59792-2_1

Cabral, A. 1974. "National Liberation and Culture." *Transition* 45: 12–17.

Carino, J. 2009. "Poverty and Well-being." In *State of the World's Indigenous Peoples (SOWIP)*, 13–50. New York: United Nations Department of Economic and Social Affairs.

Crawhall, N. 2016. "An Introduction to Integrating African Indigenous & Traditional Knowledge in National Adaptation Plans, Programmes of Action, Platforms and Policies." *Indigenous Peoples Coordinating Council of Africa (IPACC)*. https://ipacc.org.za/wp-content/uploads/2020/02/LimaReportFinal.pdf

Dalilah, Z. 2022a. "Dissecting Success: Shell vs the Wild Coast and How the Corporation was Stopped (for now)." *Africans in Diaspora*. February 8. https://www.africansinthediaspora.org/blog/shellvwildcoast

Dalilah, Z. 2022b. "What Right do we have to Avoid Risk When Movement Leaders Face Death for Daring to Make Change?" *Africans in the Diaspora*, March 28. https://www.africansinthediaspora.org/blog/assassinationriskandphilanthropy

Davis, M. 2006. *Planet of Slums*. London and New York: Verso.

FAO, AUC, ECA and WFP. 2023. Africa—Regional Overview of Food Security and Nutrition 2023:Statistics and trends. Accra, FAO. https://doi.org/10.4060/cc8743en. https://openknowledge.fao.org/server/api/core/bitstreams/c6c81d5f-e337-4b3e-8330-555c9ed0e741/content

Fanon, F. 1967. *Black Skins, White Makes.* New York: Grove Press.

Frame, M. 2022. "Ecological Imperialism: A World-Systems Approach." *American Journal of Economics and Sociology* 81(5): 503–34. https://doi.org/10.1111/ajes.12472

Ganz, G. 2020. *The Housing Crisis in Sub-Saharan African Slums.* The Borgen Project. https://borgenproject.org/sub-saharan-african-slums/

Gebremikael, M. B. 2019. "Nigusi Memarta Afari Mebata: The King and his Rules Shall Pass, but the Ways of the Afar Shall Last Forever." *Alliance for Food Sovereignty in Africa (AFSA).*

Goldtooth, T. 2019. *"Nature-Based Solutions" Greenwash Pollution. Skyprotector Briefing # 13.* Indigenous Environmental Network. https://skyprotector.org/wp-content/uploads/2019/11/Sky-Protector-13-Nature-Based-Solutions.pdf

Hare, N. 1970. "Black Ecology." *The Black Scholar.* 1(6): 2–8. https://www.jstor.org/stable/41163443

Hubert, C. 2019. *The United Nations Declaration on the Rights of Peasants: A Tool in the Struggle for a Common Future.* CETIM. https://www.cetim.ch/wp-content/uploads/The-UN-Declaration-on-the-Rights-of-Peasants.pdf

Indigenous Peoples of Africa Coordinating Council (IPACC). 2021. *Workshop on Integrating Indigenous and Traditional Knowledge in National Adaptation Plans and Strategies, Facilitators Guide.* https://www.ipacc.org.za/wp-content/uploads/2021/11/English.pdf

Indigenous Peoples of Africa Coordinating Council (IPACC). 2023. *Green Financing, a Just Transition to Protect Indigenous Peoples' Rights?.*

Jansen. L. 2020. *Indigenous World 2020—South Africa.* IWGIA. https://www.academia.edu/44873282/Indigenous_World_2020_South_Africa

La Via Campesina. 1996. *Via Campesina: The Right to Produce and Access Land.* La Via Campesina. http://safsc.org.za/wp-content/uploads/2015/09/1996- Declaration-of-Food-Sovereignty.pdf

La Via Campesina. 2007a. *The Declaration of Nyeleni.* https://nyeleni.org/spip.php?article290

La Via Campesina. 2007b, November 10. "Small Scale Farmers are Cooling Down the Earth." *La Via Campesina Views* 5: 1–24. https://viacampesina.org/en/small-scale-sustainable-farmers-are-cooling-down- the-earth/

La Via Campesina. 2008, January. *Food sovereignty for Africa: A Challenge at Fingertips* (Compilation of documents from Nyeleni La Via Campesina Conference, February 23–27, 2007). http://docplayer.net/24984357-Food-sovereignty-for-africa-a-challenge-at-fingertips.html

La Via Campesina. 2013, July 11. *From Maputo to Jakarta: 5 Years of Agroecology in La Via Campesina.* La Via Campesina Jakarta. https://viacampesina.org/en/from-maputo-to-jakarta-5-years-of-agroecology- in-la-via-campesina/

Mbembe, A. 2003. "Necropolitics." Translated by L. Meintjies. *Public Culture* 15(1): 11–40.

Mbembe, A. 2019. *Necropolitics* (S. Corcoran, trans.). Durham and London: Duke University Press.

McKeon, N., M. Watts, and W. Wolford. 2004. *Peasant Associations in Theory and Practice.* UNRIDS.

McKittrick, K. 2011. "On Plantations, Prisons, and a Black Sense of Place." *Social and Cultural Geography* 12(8): 947–63. https://doi.org/10.1080/14649365.2011.624280

Meadows, D. H., Meadows, D. L, Randers, J. and Behrens III, W. W. 1972. *The Limits to Growth: A Report of the Club of Rome's Project on the Predicament of Mankind.* New York: Universe Books.

Mochire, D., Losale, C., Mbelu, A., Rokatoveo, F. and Ilunga, J. 2022. "Democratic Republic of the Congo." In D. Mamo (ed.), *The Indigenous World 2022,* 67–74. International Working Group on Indigenous Affairs.

Moyo, S. 2013. "Land Reform and Redistribution in Zimbabwe Since 1980." In S. Moyo, and W. Chambati. (eds.), *Land and Agrarian Reform in Zimbabwe: Beyond White Settler Capitalism,* 29–78. CODESRIA.

Natural Justice. 2019. The Traditional Knowledge of the Khoikoi and San Acknowledged through the Benefit Sharing Agreement for Rooibos. https://naturaljustice.org/the-traditional-knowledge-of-the-khoikhoi-and-san-acknowledged-through-the-benefit-sharing-agreement-for-rooibos/

Ngwane, T. 2017. "Against All Odds: The Spirit of Marikana and the Resurgence of the Working-Class Movement in South Africa." In T. Ngwane, L. Sinwell, and I. Ness. (eds.), *Urban Revolt: State Power and the Rise of People's Movements in the Global South.* Chicago: Haymarket Books.

Ngwane, T. 2021. *Amakomiti: Grassroots Democracy in South Africa's Shack Settlements.* London: Pluto Press.

Nixon, R. 2011. *Slow Violence and the Environmentalism of the Poor.* Massachusetts: Harvard University Press.

No REDD in Africa Network. 2015. *Stopping the Continent Grab and the REDD-ification of Africa.* No REDD in Africa Network. https://no-redd.com/wp-content/uploads/2016/11/REDDinAfricaNetwork.pdf

Pain, R. 2018. "Chronic Urban Trauma: The Slow Violence of Housing Dispossession." *Urban Studies* 56(2): 385–400. https://doi.org/10.1177/0042089018795796.

Pinto, Jr., L. 2019. *Sustaining the traditional production and conservation of Creole rice seeds in Guinea Bissau.* Alliance for Food Sovereignty in Africa (AFSA).

Rabaka, R. 2009. *Africana Critical Theory: Reconstructing the Black Radical Tradition, from W. E. B. Du Bois and C. L. R. James to Frantz Fanon and Amilcar Cabral.* Lanham, MD: Lexington Books.

Rabaka, R. 2010a. *Against Epistemic Apartheid: W. E. B. Du Bois and the Disciplinary Decadence of Sociology.* Lanham, MD: Lexington Books.

Rabaka, R. 2010b. *Forms of Fanonism: Frantz Fanon's Critical Theory and the Dialectics of Decolonialization.* Lanham, MD: Lexington Books.

Rabaka, R. 2014. *Concepts of Cabralism: Amilcar Cabral and Africana Critical Theory.* London: Lexington Books.

Rabaka, R., Ed. 2020. *Routledge Handbook of Pan Africanism.* London: Routledge.

Rainforest Foundation. 2009. The Situation of the Forest Peoples in the Central African Republic. Rainforest Foundation. https://www.rainforestfoundationuk.org/media .ashx/thesituationofforestpeoplesofcar.pdf

Rodney, W. 1975. "Aspects of the International Class Struggle in Africa, the Caribbean and America." In H. Campbell (ed.), *Pan-Africanism: Struggle, Neo-colonialism, and Imperialism—Documents of the Sixth Pan African Congress* 18–41. Afro-Carib Publications.

Rural Women's Assembly. 2015a. Rural Women's Assembly Newsletter #1. https:// ruralwomensassembly.files.wordpress.com/2015/10/rwa-newsletter-040215-pdf-2–1. pdf

Rural Women's Assembly. 2015b. Rural Women's Assembly Newsletter # 2. https:// ruralwomensassembly.files.wordpress.com/2015/10/rwa-newsletter-2-final-2.pdf

Sadomba, Z. W. 2013. "A Decade of Zimbabwe's Land Revolution: The Politics of the War Veteran Guard." In S. Moyo and W. Chambati. (eds.), *Land and Agrarian Reform in Zimbabwe: Beyond White Settler Capitalism*, 79–122. CODESRIA.

Sena, K. 2015. *Indigenous Defenders of Africa: A Funders Brief.* Indigenous Peoples of Africa Coordinating Committee.

Sinwell, L. 2017. "Thembelihle Burning, Hope Rising." In T. Ngwane, L. Sinwell, and I. Ness. (eds.), *Urban Revolt: State Power and the Rise of People's Movements in the Global South*, 15–28. Haymarket Books.

United Nations Department of Economic and Social Affairs. 2019. *State of the World's Indigenous Peoples: Implementing the United Nations Declaration on the Rights of Indigenous Peoples.*

United Nations Human Rights Council. 2018. *United Nations Declaration on the Rights of Peasants.* United Nations General Assembly. https://www.geneva-academy.ch/ joomlatools-files/docman-files/UN%20Declaration%20on%20the%20rights%20of %20peasants.pdf

United Nations Permanent Forum on Indigenous Issues. 2013. Indigenous Peoples in the African Region. Backgrounder for Twelfth Session of the UN Permanent Forum on Indigenous Issues, May 23. https://www.un.org/esa/socdev/unpfii/documents /2013/Media/Fact%20Sheet_Africa_%20UNPFII-12.pdf

US. AID. 2021. Indigenous Peoples Regional Profile: Sub-Saharan Africa. https://www .usaid.gov/sites/default/files/2022-05/IP-Regional-Profile-Sub-Saharan-Africa.pdf

10

Racial Capitalism, Imperialism, Eco-Apartheid, and the Necessity of Fundamental Social Change

Rose M. Brewer

Introduction

This chapter is an analysis of racial capitalism, eco-apartheid, and more precisely, *imperial racial monopoly capitalism* and *global eco-apartheid*. This era of neoliberalism and intense privatization is devastating for people and the planet, featuring a fundamentally imperialist global capitalist system. Human and ecological destruction are the consequences of this rabid, dystopian order. The Global South in particular bears the heavy dispossessions of eco-imperialism, extractivism, and the human/nature divide catalyzed by this system. The 500-year legacy of white supremacy, colonialism, capitalism, and imperialism are the deep structures of the current order, articulating the divide between humanity and nature (Amin, 2019). Billions of the world's peoples are excluded from water, food, education, housing, and a living wage. These are the consequences of exploitation in the never-ending search for profit, while the resources of the Earth, which should be shared and tended, are in the hands of a tiny global elite.

As has always been the case historically, the fundamental logic of twenty-first-century imperial racial monopoly capitalism is the accumulation of capital. Today, this system faces multiple crises. Capitalism's imperialist dimension catalyzes eco-apartheid through the expropriation of resources from the periphery. Thus, eco-apartheid looms large and is embedded in a global racial logic intertwined with economic dispossession and ecological destruction. Given this, the consequences are dire for the world's peoples and especially for the African world in the United States, the African continent, and across the diaspora (Rodney, 1981). While eco-apartheid, austerity, extractivism, race, gender, and class are interrogated in this chapter, so is Black women-led resistance(s) (Mersha, 2018). These struggles are core to the fight against eco-apartheid and global imperialism. These are struggles of visioning, creation, and practice for a new social order for people and the planet. Building global solidarity through deep organizing across national boundaries is the imperative for global social transformation.

Where to Begin?

Understanding the positioning of the Black world, given the focus on imperial racial monopoly capitalism, and what that means for centering the environmental imperative is complicated. Such an analysis involves making sense of a number of complexities and contradictions. At the center of this complexity is my contention that one of the most powerful revolutionary visions of Black life across Africa and the African diaspora is the defeat of white world supremacy and racial monopoly capitalism. This is the struggle for self-determination, given Europe's underdevelopment of Africa (Rodney, 1981). Pan-Africanism is a historical political mechanism to achieve this goal across Africa and the African diaspora (Rabaka, 2020). As the climate catastrophe traverses the Global South, creating ecocide and eco-apartheid, whether in the cancer alleys of the US South or an African continent on fire, the struggle for social transformation increasingly has the environmental imperative as a core site of struggle.

From the fight against colonialism and imperialism through the drive for Black Power in the twentieth century, to the Black Lives Matter movement today, we are in a period of uprisings for fundamental social change. At the core of the contemporary fight in the African world is the struggle against eco-apartheid and imperial global capitalism. Indeed, those peoples least responsible for environmental catastrophe are paying the heaviest price given imperial racial monopoly capitalist extractivism and neocolonialism (Bassey, 2020). This global order drives environmental devastation and catalyzes the demand for self-emancipation and self-determination.

In this very difficult period of resurgent white nationalism, police murders, drought, floods, fires, hunger, and state violence, there is deep, though uneven, resistance to these realities. Increasingly core to this period is the uprising against climate disaster, capitalist extractivism, and environmental devastation. These are the key imperatives of change to transform the current order. Across Africa and the African diaspora, this uprising occurs in the context of the expropriation of land, labor, minerals, desertification, and a planet on fire—death. Yes, imperial racial monopoly capitalism articulates a world-system where the poorest people on the face of the Earth are Black, Brown, Indigenous, young, female, and children. Human and nonhuman forms face extinction. This social reality is expressed in a set of complex social relations and political dynamics that are discussed in this analysis.

In 2022, the smoke and mirrors propagated by the chief purveyors of climate catastrophe, US multinational corporations, and imperial racial monopoly capitalist bourgeois politicians, opened another ideological front with something called the Inflation Reduction Act (2022). Investopedia (2022) notes that, on paper, the Inflation Reduction Act (IRA) will make the single largest investment in climate and energy in American history: "Spending is designed to lower energy costs, increase cleaner energy production, and reduce carbon emissions by 40% by 2030." On closer scrutiny, a number of red flags undercut the professed climate benefits of the act (Investopedia, 2022). Specifically, some provisions of the IRA actually increase fossil fuel production on public lands. For example, regarding the use of public lands, these provisions include the following:

- New requirements to hold lease sales that open up new oil and gas production
- Reinstatement of a recent offshore oil and gas lease sale that was struck down on environmental grounds
- Requirement that the Interior Department hold at least three more offshore oil and gas lease sales
- Minimum royalties increase for companies that extract oil and gas on public lands and waters
- Added royalty for public land and water extraction of gas that is later burned off or released as waste instead of sold as fuel (Investopedia, 2022).

Another tough critique of the IRA is waged by the Indigenous Environmental Network. They state:

This Act is not a climate bill. It is a Trojan Horse. By co-opting the language and demands of environmental justice and frontline communities towards environmental and climate justice, the IRA is a wholesale package to distract the public from the inaction of the federal government and ensure our path to climate catastrophe. It does not adequately address the root cause of the climate crisis and ban extractive industries from exploiting the Earth. Instead, it relies heavily on climate criminals to solve the interlinked climate and economic crisis. (Indigenous Environmental Network, n.d.)

So How Did We Get Here?

What is this system of imperial racial monopoly capitalism? LeftRoots, a cadrefication organization working with organizers to understand strategy and become strategists, asserts:

> It is a profit-driven system, simultaneously located in the logics of imperialism, white supremacy and heteropatriarchy. Monopolies exploit labor and natural resources, and the exportation of finance capital, rather than manufactured goods to sustain colonialism, which is an integral function of imperialism. The word "imperial" is used in this stage of capitalism to emphasize the global nature of the capitalist system. (LeftRoots, n.d.)

Interrogating the concept further, "monopolies" refer to the concentrated economic power of capitalist corporations. The economic power is concentrated to such a degree that a small elite controls world markets and all productive systems (Foster, 2015). Moreover, the increasing financialization of United States and global capitalism means that a growing number of the capitalist class gets its wealth from finance rather than production. Its imperialist reach is captured by Samir Amin:

> Capitalism is not only a system based on the exploitation of labor by capital. It is also a system based on polarization in its development on the world scale. Capitalism and imperialism are the two inseparable faces of the same reality. (Amin, 2019)

Further, Paschel reminds us of Walter Rodney's thesis on imperialism and how Europe underdeveloped Africa. She describes Rodney's argument thusly:

> African countries had been locked in what might be called a colonial division of labor, whereby natural resources are extracted, and act as the raw materials for industrialization in the North. This arrangement, he [Rodney] argued, brought about a specific form of underdevelopment whose logic was built on serving the needs of the metropole or core countries. (Paschel, 2016, p. 6)

Therefore, imperialism is a logic, a practice, which expropriates (takes) the world's resources (land, labor, mineral, etc.) and shifts them into the coffers of a small elite of global capitalists, heavily represented by the United States.

As V. J. Prasad articulates, "Imperialism is the underlying structure of plunder and domination, which can take place without direct control of another people's political sovereignty" (Prashad, 2023).

Moreover, imperialism is simultaneously expressed ideologically in the cultural commodification of Blackness—stereotyped and terrorized in the international division of labor as incapacity, belonging at the bottom of the global hierarchy of nations. This cultural imperialism is expressed in consciousness (Fanon, 2005). For instance, as Angela Davis (2003) observes, we know (or are conditioned to think we know) what crime looks like because it inculcates the ideology of who belongs at the bottom of the global and domestic economic order. Those at the bottom deserve to do low-paid, highly exploited labor because that's who they are. This is the long-standing racial ideology which divides the world's peoples into deserving and undeserving, inferior and superior. Nonetheless, while the face of imperialism is the European male, many of the elite beneficiaries of the periphery are Indigenous to oppressed cultures. These Indigenous elites represent a class integrated into the logic of transnational capital to the benefit of the multinationals and their own wealth (Rodney, 1981).

Under imperial racial monopoly capitalism, the Black body was transformed into an object rather than a human subject (Fanon, 2005). Colonialism and imperialism underpin global capitalist development historically and today through their accumulation logic. The enslavement of African peoples was part and parcel of this accumulation process. The enslaved saw none of the profits or social capital they produced and were subject to the deprivation of humanity and rights, rape, physical violence, and generations of dispossession and murder.

Since the fifteenth century, plunder, exploitation, and expropriation in the colonial order built the global imperial core with the majority of the world's people exploited in the periphery. Racialized slavery, settler colonialism, land theft, and genocide in the United States generated an internal periphery of superexploited Black, Brown, Indigenous, and Asian workers, whose labor and land were taken through expropriation, exploitation, and rape of land and bodies (The Red Nation, 2021). A disproportionate number of the rapable were and are women. The United States is founded on this white supremacy, land theft, genocide, and settler colonialism which underpins its emergence as a global capitalist empire.

When in the late nineteenth century the world was feeling the birth pains of industrial capitalism, African peoples too transitioned, struggling to exist in a global capitalist order that had already set in place racist, sexual, and gendered mechanisms to keep them from being treated as fully human and self-determining. In the US context, Black people increasingly moved/were pushed into urban industrial centers by the late nineteenth and early twentieth centuries. They were socially excluded but exploited as expendable cheap labor in fields, factories, and various services. Until the mid-twentieth century, Black women were almost completely locked into domestic service work in white homes.

Globally, the African world acted politically in the wake of global white supremacy and the partition of Africa. In this context, the Pan-African conferences were organized, pressing for the political unity of the African world and Black self-determination (Rabaka, 2020). W. E. B. DuBois would play a key leadership role in this process (Rabaka, 2020). Self-determination and decolonization were the hue and cry of these conferences and their political intent.

After the Second imperialist World War and the agreements made at Bretton Woods to establish the institutional arrangements to solidify the economic, political, and social structures of the global capitalist system, the World Bank and International Monetary Fund (IMF) were organized in the service of imperial global capitalism. At the heart of this international system is the never-ending quest for cheaper labor and the easy flow of capital goods, debt, and profit (Foster, 2015). This perverted form of exploitation and expropriation transplanted skilled and unskilled labor to new shores, exploiting the economic and sexual landscapes of Asia, Africa, the Caribbean, and South America.

The economic logic in the dominant Global North is mirror-imaged in the Global South through the policies of the IMF, World Bank, and private capital. Imperial racial monopoly capital thus sources ecocide and genocide through the established institutions of global capital. The driving force of this is the constant search for profit and power. Meanwhile, race and class are deeply enmeshed in gender, sexuality, and nation, simultaneously shaping and being constructed by political economy and ideology.

The capitalist world-system represents continued wealth expropriation and the destruction of the natural world—from poor to rich. Technology and finance, which include computerization, robotization, and the growth of electronics, are at the heart of its contemporary dynamics. They are the mechanisms through which wealth transfers occur and through which labor is transformed or eliminated. Foster (2015) makes the logic of financialization quite clear:

> the financialization of U.S. capitalism over the last four decades has been accompanied by a dramatic and probably long-lasting shift in the location of the capitalist class, a growing proportion of which now derives its wealth

from finance as opposed to production. Finance capitalist exercise power by controlling access to the markets through which capital accumulation becomes investment, directing flows of capital in various forms—as equity purchases, bond sales, direct investment, etc.—to places and user that are approved by the financial analytic structure of the Wall Street and City of London banks and investment firms.

Given this conjuncture, it is a fact that, in general, many sectors of the Black population in the United States, the Caribbean, Europe, Africa, and the diaspora are in social, economic, and political crises. These are the frontline/fenceline communities and societies bearing the brunt of environmental and climate catastrophes. In advanced Western capitalist societies such as the United States, the dismantling of the social wage and intense privatization characterize the current period of economic destruction, intensifying the harm of the poorest, most dispossessed, and heavily Black and Brown populations. Ideologically, white supremacy and white nationalism are on the rise.

Internally, within the United States, the power of capital and class divisions means that a capitalist elite profits from the divisions catalyzed within a working class that is gendered and raced. Nonetheless, there is a tier among those who have been excluded who are now incorporated within the US capitalist class. This includes, for example, a small wealth-holding sector of the Black population. Comparably, on a global level, a comprador class that benefits from global monopoly capital is part and parcel of the world-system. Thus, within Global South nations, there is a growing division between a wealth-holding elite and the masses of their people. In the Global South, this represents a convergence of global capitalism with a national capitalist class. Acting as agents for foreign organizations, trade, or economic and political exploitation, this class in the periphery or semi-periphery is integrated into the world-system while the masses of the people suffer. For example, in Senegal, structural adjustments have shifted resources from human needs to the private sector (Hesse, 2004). Education, social services, and health are sacrificed to the market. Societies such as these are burdened with debt while saddled with very narrow economic possibilities, concentrated in only those industries that will advance global capital. Given this, we must be very clear about this system of class exploitation but understand as well how it is deeply conditioned by racial practices and gender logics—an integrated dynamic of color, gender, sexual, and class oppression.

In sum, the system is advanced *imperial racial monopoly capitalism*. It is a profit-driven system, simultaneously located in the logics of imperialism, white supremacy, and heteropatriarchy. It is the driver of eco-apartheid. At the center of imperial racial capitalism are women whose labor is used to enrich a small economic elite but who also do the socially reproductive work of the world—cleaning, cooking and caring—via unpaid labor and super-exploitation that goes unnamed and unchecked (Mies, 2022). This socially necessary/reproductive labor is central to the labor exploitation in the international division of labor. It is a gendered division of labor. Nonetheless, technology platforms and robotization attempt to circumvent this. Through Covid-19, the critical role of social reproduction was revealed. The system hasn't been able to get around the need for care work, the reproductive labor key to human survival. The system survives if workers' lives are reproduced continuously and reliably while being replaced generationally. Food, housing, public transport, public schools, and hospitals are all ingredients of what Bhattacharya (2017) calls "life making," which socially reproduces workers and families.

This, of course, is connected to the fate of the working class as a whole, as women do the bulk of this feminized and devalued labor. Care work is devalued or unpaid ideologically because of racialized sexism. It is work disproportionately done by Black, Brown, and gender-nonconforming lower layers of the working class. This life-making work, done heavily in homes, schools, hospitals, and elder care, among others, is privatized or underfunded. This is the two-fold logic of racial monopoly capitalism. Reproductive labor today is broadly structured into the market while technology advances a new "gig economy" that also responds to domestic needs from food delivery, laundry, elder care, and beyond. The structural shift to low-paid reproductive labor by all genders for pay is a key feature of today's racial monopoly capitalism. It is also a site of struggle.

Women Resist Globally

The system does not go unnamed or unchecked. A powerful example of this fightback is the case of the Garifuna peoples in Honduras (Global Sherpa, 2022). Mersha describes the Garifuna community thusly: "These are Afro descendant and Indigenous peoples, whose ancestors were escaped Africans who were brought to St. Vincent on slave ships, as well as Indigenous Arawak peoples" (Mersha, 2018: 1425). The Garifuna have been fighting multiple threats to survive. Garifuna women have been at the forefront of many of these struggles, especially those against transnational companies who have been granted concessions by the Honduran government to drill for offshore oil near Garifuna territory (Mersha, 2018: 1425). These women are organized into the Organization of Garifuna Women in Honduras. One of the leading, if not the leading voice, is Miriam Miranda, who heads the Organización Fraternal Negra Hondureña (OFRANEH). OFRANEH is a grassroots organization that works on behalf of the Garifuna people in a "permanent struggle for their autonomy, as well as their collective social, economic, cultural, and territorial rights." Organizational unity and strategically fighting for their rights are at the heart of this resistance. They use a variety of tactics to do so, including direct action, land reclamation, legal strategies, communications/cultural resistance, and building international solidarity (Mersha, 2018: 1425).

It is through this strategic use of international solidarity that these Afro-descendant women share valuable lessons for the global emancipatory struggle. Indeed, this international movement-building holds great promise. Miriam Miranda and I have been in common space together, strategizing how to build stronger linkages between African diaspora fighters in the United States and those in Honduras. This internationalism is reciprocal with US activists traveling to Honduras and joining in the common work of OFRANEH. Solidarity is centered on standing within the circle of struggle, not outside of it.

The freedom fighters of Garifuna shed light on what needs to be done. They are unabashed in their commitment to self-determination and creating a different social order for the Garifuna and the oppressed. They represent the transformational commitment to create a different social order. It is quite a potent moment when Black, Brown, and Indigenous women see themselves as a class opposed to capitalism and stand to withhold their labor, both reproductive and productive. They are thus leading forces for social change—building political movements in opposition to the destruction of people and the planet. They know that organizing is an imperative.

Organizing Is an Imperative

As the African world continues to live out the legacies of colonialism, neocolonialism, imperialism, and the current inequalities of global capital, the internal struggle for those fighting within the imperial core is critical. A domestic example of this organizing for revolutionary change is found in the work of Cooperation Jackson in Jackson, Mississippi. Cooperation Jackson is building an ecosocialist alternative, and its work articulates the commitment to building the future in the present (Akuno, 2023). Kali Akuno, co-director of Cooperation

Jackson, describes the organization's theory and practice of social change. He asserts that the program and strategy of Cooperation Jackson is intended to accomplish four fundamental ends:

- To place ownership and control over the primary means of production directly in the hands of the Black working class of Jackson.
- To build and advance the development of the ecologically regenerative forces of production in Jackson.
- To democratically transform the political economy of the City of Jackson, the State of Mississippi, and the Southeastern Region of the United States.
- To advance the aims and objectives of the Jackson-Kush Plan, which are to attain self-determination for people of African descent and the radical, democratic transformation of the State of Mississippi (Akuno, 2023, pp. 12–13).

This vision is deeply informed by a rootedness in the local, but articulates "the radical decolonization and transformation of the United States itself" (Akuno, 2023, p. 13). It is connected to an ecosocialist transformation.

Ecosocialists such as Victor Wallis (2018) are clear about the imperative of moving from a capitalist to an ecosocialist society, which means dismantling the capitalist system. This is a necessity to liberate humanity and the Earth. Radicals are making intersecting connections between ecological transformation and other struggles against police violence and for health, housing, and education. For example, the work of Igietseme links the struggle for the abolition of police and the violence they inflict on the Black community to the environmental devastation which the Black community disproportionately experiences (Igietseme, 2014).

A number of activists are clear that those least responsible for climate catastrophe are bearing the greatest costs (Johnson, 2011). Whether we are talking about Hurricane Katrina when the levees broke, intentionally flooding the ninth ward, or burning, flooding, and extractivism on the African continent, there is a disproportionate impact on those societies least responsible for the climate crisis (Johnson, 2011; Bassey, 2020). From the cancer alleys of Louisiana to the Ogoni peoples' struggles against Shell Oil, these processes represent the reality of the 500-year legacy of imperial racial monopoly capitalism. Imperialist extractivism destroys. This extraction is culpable for decades of releasing greenhouse gasses as well as the degradation of land and water for profit (Wengraf, 2018). Once again, building an alliance of oppressed peoples across the globe is at the heart of Samir Amin's vision of social transformation (Amin, 2019).

Environmental justice groups such as Indigenous Environmental Network and the Red Nation with their *Red New Deal* place environmental justice as well as being in right relations to Mother Earth at the center of resistance. The Red Nation is also clear about the need to create a socialist alternative. Indeed, through the creation of organizations and networks, through building power for Black/Brown/Red self-determination, land, and protection, current struggles unfold. From the Garifuna fight for land back and justice in Honduras to Cooperation Jackson struggling for self-determination and ecosocialism, what must be done is clear.

In short, the hue and cry is that we must struggle for a new society. But the fight back, the social transformation, must encompass a vision untried in human history: struggling against these oppressions in deep relationality to one another, centering human and nonhuman, nature and Mother Earth, as interconnected—with a deep clarity of the logic of imperial racial monopoly capitalism.

Essentially, the charge for the African world and other oppressed peoples is strategically organizing highly difficult and critical movement-building efforts in deep relationships with committed radicals and revolutionaries. This entails connecting forces in motion that are fighting against the capitalist system and its imperialistic face. Samuel Grant, Black environmentalist and co-editor of this volume, sums up possibilities emerging from Africa and the African diaspora thusly:

> African Nationalists propose the rights of states and a strengthening of sovereignty over natural resources, labor, monetary policy, and the environment within their borders. Internationalist-Anti-Corporate Globalization movements propose a dismantling of the corporate-First World political economic pact that ushers in rapid privatization and liberalization. Poor peoples movements, whether urban or rural, make claims alternately on local authorities, the state and on global intermediaries to assert their rights just to survive.

We know much organizing work remains to be done. Our movements must act to end this highly unequal global capitalist system. While difficult, this stark moment of dispossession, climate catastrophe, and travesty requires us to lead with complex theorizing and practice (Brewer, 2010). This is the change that must occur for another world to be born. At the core of international solidarity is building linked struggles across the globe. This means connecting through communication, sharing resources, and deep engagement with unity-struggle-unity, the tried and true method for building transformational change and global social movements.

The gamble is on by elites that stringent policies can be imposed and that war, militarism, and violence will win the day without organized resistance. The ideology of the right, coercion, ideological attacks and state violence are their tools. But the historical record speaks: that as the world's peoples, we will resist, rebel, and transform.

References

Akuno, Kali. 2023. "Build and Fight: The Program and Strategy of Cooperation Jackson." In K. Acuno and M. Meyer (eds.), *Jackson Rising Redux: Lessons on Building the Future in the Present*, 12–12. Oakland, CA: PM Press.

Amin, Samir. 2019. "The New Imperialist Structure." *Monthly Review* 71(3): 13. Retrieved from https://monthyreview.org/2019/07/01/toward-the-formation-of-a-transnational-alliance-of-working-and-oppressed-peoples

Bhattacharya, Tithi, Ed. 2017. *Social Reproduction Theory: Remapping Class, Recentering Oppression*. London: Pluto Press.

Bassey, Nnimmo. 2020. *To Cook a Continent: Destructive Extraction and Climate Crisis In Africa*. Oxford: Pambazuka Press.

Brewer, Rose. 2010. "The Social Forum Process and the Praxis of Race, Class Gender, Sexualities." In R. Dello Bueno and D. Fasenfast (eds.), *Social Change, Resistances, and Social Practices*, 57–62. Leiden, The Netherlands: Brill Publishers.

Davis, Angela. 2003. *Are Prisons Obsolete*. New York: Seven Stories Press.

Fanon, Frantz. 2005. *The Wretched of the Earth*. New York: Grove Press.

Foster, John B. 2015. "The New Imperialism of Globalized Monopoly-Finance Capitalism." *Monthly Review*. Retrieved from https://monthlyreview.org/2015/07/01/the-new-imperialism-of-globalized-monopoly-finance-capital

Global Sherpa. 2022. "The Garifuna People, History and Culture." Retrieved from globalsherpa.org.

Hesse, Brian. 2004. "The Peugeot and the Baobab: Islam Structural Adjustment and Liberalism in Senegal." *Journal of Contemporary African Studies* 22(1): 3–12.

Igietseme, Nene. 2014. "Black Organizing, and Mike Brown. Center for Story Based Strategy." Retrieved from https://www.storybasedstrategy.org/blog-full/2017/11/29/climate-justice-black-organizing-and-mike-brown

Indigenous Environmental Network. n.d. "The Inflation Reduction Act of 2022 is Not a Climate Bill." Retrieved from https://www.ienearth.org

Inflation Reduction Act. 2022. Retrieved from https://wwww.whitehouse.gov

Investopedia. 2022. "Inflation Reduction Act." Retrieved from https://www.investopedia.com/inflation-reduction-act-of-2022-636226

Johnson, Cedric, Ed. 2011. *The Neoliberal Deluge: Hurricane Katrina, Late Capitalism, and the Remaking of New Orleans.* Minneapolis: University of Minnesota Press.

LeftRoots. n.d.. "About." Retrieved from https://leftroots.net

Mersha, Sara. 2018. "Black Lives and Climate Justice: Courage and Power in Defending Communities and Mother Earth." *Third World Quarterly* 39(7): 1421–34.

Mies, Maria. 2022. *Patriarchy and Accumulation on a World Scale.* New York: Bloomsbury Press.

Paschel, Tianna. 2016. "Walter Rodney and the Racial Underpinnings of Global Inequality." New York: Social Science Research Council. Retrieved from https://items.ssrc.org/what-is-inequality/walter-rodney-and-the-racial-underpinnings-of-global-inequality/

Prashad, Vijay. 2023. "Imperialism and the Rise and Fall of US Unipolarity." Retrieved https://youtu.be/VleW3S-xh8U

Rabaka, Reiland, Ed. 2020. Ed. *Routledge Handbook of Pan Africanism.* Oxfordshire: Routledge.

The Red Nation. 2021. *The Red Deal: Indigenous Action to Save Our Earth.* New York: Common Notions.

Walter, Rodney. 1981. *How Europe Underdeveloped Africa.* Washington, DC: Howard University Press.

Wallis, Victor. 2018. *Red-Green Revolution: The Politics and Technology of EcoSocialism.* Toronto: Political Animal Press.

Wengraf, Lee. 2018. *Extracting Profit, Imperialism, Neoliberalism, and the new Scramble for Africa.* Chicago: Haymarket Books.

11

Estranged in Urban Beirut

Anthropocentric Ecological Destruction, Mysticist Shia Islam, and Epistemological Alternatives

Ali Kassem

Over the past decades, attention to envisioning alternative civilizational models to Eurocentric modernity/coloniality has grown. Within this, the ecological challenge is central in the face of an ongoing modern anthropocentric destructive consumption of earth and life. Accordingly, much scholarship has sought to work alongside various Indigenous communities across the globe to learn and rethink practices and knowledges. (World) "Religions," particularly "Abrahamic religions," have figured much less within this. Often, there even lingers an underlying assumption that these religious systems are at the core of the current anthropocentric crisis. This chapter grates against this assumption as it presents lived Muslim alternative cosmologies and modes of being in the world incommensurable with anthropocentric modernity/coloniality and with substantial potential to disrupt its ecological violence and assault.

This argument is based on qualitative research conducted between 2019 and 2022 exploring lived experiences, discrimination, urban living, and urban-rural connectivity within Muslim communities in Lebanon. Within this research, a powerful theme of alienation and estrangement from nature emerged in the country's urban spaces, particularly the capital city of Beirut. Discussing this with participants, a (small) group of them articulated an incommensurability between their Muslim understandings and modern urban life's fracturing away and destruction of earth and nature. This chapter presents this critique—a specifically Mysticist Shia one—as it indicates its potentials in moving beyond anthropocentrism.

Lebanon, "Islam," and Methods

Lebanon is a small Eastern Mediterranean country that "never existed before in history. It is a product of the Franco-British colonial partition of the Middle East" (Traboulsi, 2007: 75). It is a religiously diverse country with over nineteen official religious sects, and its population is divided between various Christian denominations as well as multiple Muslim ones. A key pillar of Lebanon's foundation in the early half of the twentieth century was its projected role as a beacon of the French *mission civiliatrice* in the Arab East (Traboulsi, 2007; Salibi, 1998) as part of Eurocentric modernity's larger project of domination and hegemony in the region. Economically, a capitalist, liberal, laissez-faire model at the service of a "westernised" ruling elite has long been dominant—arguably since Lebanon's very colonial creation (Kardahji, 2015). With finance, trade, and services made central, the country was birthed as—and continues to be—a nation of deep inequalities and instabilities. In more recent decades, the international neoliberalizing capitalist institutions of IMF and World Bank have been key actors in this respect (Traboulsi, 2007; Baumann, 2017).

In terms of geography, large coastal cities have long stood in stark contrast to much of Lebanon's interior, hinterland, and border regions which are unequally and systematically "under-developed." Meanwhile, Lebanon's major cities—where the overwhelming majority of the population resides—were made into key metropolitan, economic, and touristic sites of the Arab-majority world (Hamade, 2014: 56; Traboulsi, 2007). Consequently, the country itself, deeply centralized in Beirut politically, economically, and culturally, is under an urban hegemony as non-urban or semi-urban spaces continue to be systematically impoverished, neglected, and often excluded from the national economy (Fawaz, 2009).

Over the past decades, Beirut has continued to be reconfigured toward further urbanization and "development" (Makarem, 2014; Fawaz, 2009; Krijnen and Fawaz, 2010). This includes the privatization of most, if not all, public spaces in the city and the ongoing gradual selling-off of remaining parks or green spaces to private investors and multinational corporations for various "developmental" projects—from shopping malls to private luxury condominiums. In this respect, Cherkawi (2020: 38) identifies "neoliberal public-private sector partnership" as the key agent paving the way for "urban neoliberal projects" that have powerfully led to the continuous vanishing of already scarce green spaces—particularly public ones.

The country today faces serious ecological crises. These include an ongoing waste crisis, deforestation, food insecurity, the contamination of much of its drinking water, and one of the highest rates of air pollution in the region. These crises are, it is key to note, all deeply entwined with Lebanon's persistent neoliberal developmentalism under IMF and World Bank patronage as well as its history of postcolonial conflict and political dysfunction (Khodr and Awwad, 2021).

Across public, media, and academic debates, solutions to Lebanon's ecological crises are systematically envisioned within the confines of advanced "technological fixes," individualized consumer behaviors such as reuse and recycling, and alignment with "international" (UN and EU) environmental policies. Within this, many have called for the empowerment of the Lebanese state and its defense against corruption as a key avenue of resolving such crises. Redress and alternatives beyond the realm of technological modernization or state-centrism remain largely absent in Lebanon, as in the wider region.

This chapter is informed by ethnographic and qualitative research conducted specifically with Lebanese Muslim Shias in Lebanon between May 2019 and May 2021. It is supported by follow-up interviews conducted between March and May 2022. The chapter especially builds on conversations with five students of Islamic studies and one Muslim environmental activist residing in greater Beirut with whom in-depth interviews were held. This group of participants is formed of young people educated in Lebanon's westernized universities and living in Beirut. With one exception, they were all middle or upper middle class. Their arguments, conceptions, challenges, and alternatives are not here posited as mainstream or dominant within wider Shia or Muslim societies, in Lebanon or beyond, but rather as present, even if at the margin.

Maldonado-Torres (2008: 379) argues for an engagement with "religious traditions" as possible sites of "critique and liberation" and as "repositories of knowledge and sources of theory." Surely, in doing this, the multiplicity and complexity of these religious systems, and both their oppressive and liberatory dimensions, must remain centered. Informed by this, this chapter specifically engages Islam as a contemporary everyday lived experience that manifests and articulates differently in different times and spaces and in relation to various socioeconomic conditions and questions to examine the critiques and liberations it could offer specifically in relation to anthropocentrism and contemporary ecological crises. This is in line with Asad's "anthropology of Islam" and away from any essentializing arguments, foreclosures, or claims of representation of an imagined clearly defined, complete, inherent, or self-contained "religious system."

From Estrangement to Alternative Epistemologies: Thinking in Urban Beirut

Estrangement in Urban Beirut

During conversations with this project's participants, life in urban Beirut, as well as Lebanon's other major cities, was often presented in negative and adverse terms. From high pollution and resulting diseases such as cancer to the lack of "public spaces," Beirut was experienced as a "dead" and "dangerous" place by a significant number of participants. The urban, in this sense, was drawn as a space where one *has* to live for education or employment rather than a place one would *want* to live. Key in this negative conceptualization was a significant sense of estrangement from "nature" and "earth" whereby a number of participants drew-out the urban space as one that has been fractured away. Fadi, a young engineer, expressed:

> The pollution here, the noise, the cars, it's known, Beirut is a cement fortress, there are no green spaces whatsoever. It's very sad.

In presenting these narratives about Beirut and what I would term its estrangement from nature, many participants pitted this as "un-Islamic," even as a mode of life that made the Islamic impossible. Mohammad, an older man living in Dahieh, Beirut's southern suburb, expressed as he discussed the absence of green spaces in Beirut:

> You see, in the Quran, God talks about nature all the time. The second most repeated word in the Quran, after the word God, is earth. That has to say something. All over, God tells us to contemplate nature and reflect. But now, living in Beirut, what is there to contemplate? What nature? We are so distanced from it, it just makes it impossible. It's not in our everyday [lived experience] anymore.

Mohammad's understanding of urban life reflected a sense that it made the "Muslim impossible" as the city's estrangement from nature enforced residents' estrangement from nature. Mansour, a young student, turned to *Hadith*— narrations attributed to the Islamic prophet considered as authoritative sources of an Islamic belief system and way of life—to similarly express:

> We have all these hadiths, about the prophet and nature, or the prophet and animals. For example, there's a hadith that the prophet was once somewhere, and a camel came close by, and started talking to the prophet, and they had a conversation, and then the prophet called for the camel's owner and reprimanded

him, he said the camel had complained about not being treated well enough and the prophet was very troubled by this. He made a big deal of it with the man and his tribe. There are so many hadiths like this, about trees too and how the prophet used to speak to trees and how they loved him. This is all over our traditions, but no one talks about it because people will think we're a crazy religion.

Mansour went on to explain that religion has to be "rational" to appeal to people today and that Muslim scholars choose not to draw on traditions that seem "unscientific" or "unreasonable." Hadiths or stories where the prophet speaks to trees, mountains, or various animals are accordingly sidelined in favor of a reading of Islam that centers (Greek) philosophical rationality and "scientific" thinking whereby the material world is transformed into a "lifeless object." For many participants, centering such hadiths would "make the prophet look crazy." An internalized epistemicide of sorts, the hegemony of modern frames of thinking, and discourses of being, filters into Muslim subjectivity and then into religious knowledge, discourses, and practices.

Offering a similar analysis of why this component of the Muslim tradition is seldom centered on or engaged in contemporary mainstream Muslim discourses, Ousama argued:

> We rarely have these conversations, among ourselves, even though it's such an important part of our religion. And if one looks into our hadith and traditions, you will find so much about planting, about fruits and trees and seeds, about different animals, about water.

Ousama, a young student in the Hawza [Shia religious educational establishment], maintained that this is a part of the larger problem of "the day and age" where people are no longer "even aware" of what they are doing in their everyday lives. He spoke of how a particular hierarchy of professions exists in the modern world. Meanwhile, the prophets worked according to a very different sense whereby "being a farmer or a shepherd" were not only highly esteemed occupations but occupations that all prophets performed as a part of their spiritual becoming and learning. Ousama here developed a critique of the hegemonic script for "success" in the modern world, formed by specific aspirations with specific sorts of jobs that, ultimately, push people out of villages and into cities, away from their "roots and families and nature." This "inevitable" life in the city, Ousama explained, is "artificial," where even a sense of "night and day are lost." This, he argued, goes against what he presented as a "Muslim lifestyle" and pattern of everyday life where one sleeps with sundown, wakes up before sunrise, grows their own food, and develops a balanced spiritual and social life with those around them. Ousama's argument can be further identified here by looking at core Muslim rituals, including the five daily prayers organized around the movement of the sun and the annual fasting organized around the movement of the moon. Through such rituals and practices, participants developed arguments of being "synchronised to the universe" as a condition and path of "finding synchrony with God."

In discussing lifestyles, Mona, a young environmental activist living in a marginalized region of Beirut, similarly explained that Islam called for a lifestyle "in tune with nature and the natural order" rather than a "fake one like what we have in cities here." For her, that is why people continue to be so attracted to villages where they spend their weekends and summers. She argued that there was "much to be done as Islamic environmentalism, but very few people talk about it unfortunately." Positing that "the prophet had environmentalism long before the west . . . including in diets," she said:

> We're not vegetarians, not a vegetarian religion. We eat meat. But we only eat *halal* meat. There's a huge distinction there. There are only a few animals we are allowed to eat, all other animals are not to eat. And these few animals, there's a whole ritual around how they are killed. Like them being given drinking water before. Or them not seeing any other animal dying. Or not being made to feel any pain. If they do, they can no longer be eaten. Now you get everyone trying to justify this by referring to science, that their bodies produce things [chemicals] that make them unhealthy [if they get scared]. But no! It's about the ritual, it's about the meaning, it's about respect!

In a similar vein, Mona explained that the Islamic traditions discourage the eating of meat regularly, with many hadiths explicitly stating that meat is best eaten "around once a month." Literature about the prophet's lifestyle, she explained, indicated that he used to go on for about forty days without any meat. Hadith, she argued, says that this is linked to meat's "hardening of one's 'heart'" that pushes one away from the divine and spirituality. In this, the ritualism the prophet had around eating more broadly, participants argued, had been lost. Instead, a global meat industry and consumer market have taken over, again alienating them in their everyday lived experiences from what they understood to be the desired Muslim lifestyle and world order. Participants accordingly argued that a "Muslim environmentalism" held liberating and revolutionary possibilities to refuse, undo, and move beyond the contemporary order. In this respect, they found an antidote to the crisis they encountered in their everyday lives from within the Muslim standpoint. Aware of the modern animal industrial complex and its devastating effects and implication in global environmental crises (Nibert, 2012; Weiss, 2013), the possibilities herein appear particularly promising.

Fadi, a young student at the Hawza in Beirut, lamented how "we have no connection to what we eat" and contrasted this to the mode of life of the "mystics and seekers" in the past as well as "what remains of them today." He went on to explain:

> Like you know, those stories that everyone knows but no one takes seriously. Like that story of the woman who gave a thirsty cat water and then was granted heaven just for that. Or how the Imams used to share their food with any animal that came by as they ate, saying they cannot bear themselves eating while the animal looked, that they would be ashamed of the animal if they did.

Across Muslim societies, these stories are well-known. Standardly, they are presented as "ethics" and "being nice" and taught to children to encourage a particular "kindness" and "care" toward animals. Rarely are they connected to a larger cosmological understanding of the world and the place of the nonhuman within it. Yet, as participants were increasingly disillusioned with their urban life, a turn to such cosmologies seemed increasingly potent and the possibility of an ecological alternative from within such traditions developed:

> I, personally, never gave any attention to nature until I went to hajj [Muslim pilgrimage to Mecca]. In hajj, there are rules, while you are there, you cannot kill any animal no matter what. Even if a fly comes and tries to suck the blood out of you, you cannot kill her. Or step on an ant, you cannot. If you do, your whole *ihram* [state of days-long ritual worship] is voided, the whole thing! And trust me, that's big. You can't cut off a tree, you can't even walk on plants, if you do and they break or die, your *ihram* is voided. But where is that [conversation]? We're not talking about that.

Bashir explained that when he was doing hajj and he came to know all these rules, he started thinking about "the why." From there, he discovered a plethora of "an entire system that really approached nature and earth and plants and animals in a way that is so different from how we do today, us as Muslims." This, he argued, allowed him to realize the "sorry state" of modern urban life and how "sad we are"—as well as the potential to carve out an alternative to this state from within the Islamic. Moussa explained that he had a similar realization while reading the Quran, and finding that God "swears by plants and animals, we swear by God, and God swears by plants and animals, and I started thinking, how great are they for God to be swearing by them?" With such insights and positions, Muslim discursive traditions appeared to hold significant promise in challenging and pushing beyond anthropocentric modernity/coloniality—even if this potential seldom manifests. In the following sub-section, thinking alongside participants, I will attempt to elaborate on this potential.

Alternative Cosmologies and Relationalities

In asking participants how they understood and made sense of this "different" relationship to nature, earth, plants, and animals, many referred to what they termed "Islamic philosophy." Philosophy here denoted a belief system and worldview rather than a specific mode of thinking or institutionalized scholarly form of knowledge. While some said they were not sufficiently versed in this "philosophy" to elaborate and suggested the names of scholars I could read or watch on YouTube, others, and particularly the Shia Hawza students, expressed significant knowledge of such traditions. Specifically, most referred to a philosophy known, in English, as "unity of existence." This philosophy is based on a certain Mulla Sadra, a Persian Muslim mystic and theologian of the seventeenth century (Rustom, 2012). It then goes back to the *Ishraqi* school of philosophy that is often attributed to Suhrawardi—a twelfth-century Persian mystic and theologian/philosopher. In this philosophical system, the key concept is one of "unity" in line with Islamic *tawhid* or *Oneness*.

In this *unity*, my participants explained, "everything is God and God is everything, only in different manifestations and at different gradians, including us." In this, everything that exists is also *alive, conscious,* and *rights-bearing*. Participants insisted the material world to be, therefore, a "manifestation" and "appearance" of God and the divine (which are not separate from it) within which the human exists "as one among many manifestations." Everything, they confirmed, "was alive" and "in sync and balance." Most other beings were in complete harmony with this universe, while the human needed to find that harmony and integrate within it by *choice* and *movement*. This was, as I understood it, a deficit in the human that does not apply to animals or plants. They explained that this system encompasses an understanding of a human journey—what Sadra refers to as the *four travels*—that allows one to reach the purpose of their being in total unity with the world and all that is in it. For participants, "life is life," and this life and the "afterlife" are a continuity, one mesh, where death took on a different meaning. Within this, existence in this world was one stage within a larger journey that is the "the purpose of being" to reach what participants expressed as "annihilation in existence." Each person's being, their *I* in a way, would here be diluted (annihilated, in their words) into the wider being of the cosmos.

This "philosophy" appeared strongly incommensurate with binaries of sacred and profane, or material and immaterial, or heavenly and worldly—nor with Latin Christendom's conceptions of God or even of learning and becoming.[1] It was further incompatible with the debate opposing transcendence to immanence, as it existed outside discourses conceptualizing "this life" as a journey toward a *separate* "afterlife." Interestingly, the concept of rights emerged and reemerged in these conversations, albeit in an articulation distinct from that of the modern liberal subject.

In explaining this, many participants offered a number of stories of mystics who had "transcended" into a stage where they "heard," "saw," and experienced these philosophical realities throughout their lives. Hearing the walls and the mountains do *tasbih* [ritual recitation] to praying in oneness with birds and ants, to being *scattered* and fusing with the wind, and to feeling the physical and psychological pains of human and nonhuman others as if they were one's own, participants insisted that these "philosophical realities" were not philosophy in the sense of the theoretical or the abstract. They were, rather, the "deeper realities" of this world that one must and can realize and be.

Through this understanding, they made sense of Quranic verses where mountains, trees, and animals are addressed and spoken to by God as well as the prophets, and where they in turn speak back and converse. In this, these are presented as cognizant beings that are, in many respects, "wiser" and more "aware" than humans—offered as an exemplar for humans to learn from. Yet, this "journey" had been increasingly made impossible. Hence, Moussa expressed, "you wouldn't find any seekers ['mystic'] in a place like Beirut, not even in [other parts of] Lebanon, you need to go somewhere remote and peaceful, they would never live here." For him, they *could* never survive in a modern urban space as that space would make their journeys impossible. The journey, the purpose and meaning of one's being, was destroyed by urban modern living. Either way, the imperative to follow in the footsteps of such prophets and seekers to achieve this arduous task of returning to sync with the universe remained as an ideal and horizon that everyone can and should actively seek.

In understanding this, one can turn to Mahmood (2009), who had explained that the "Muslim's relationship to the Muslim prophet is predicated not so much

---

[1] Christianity is surely more complex and plural than mainstream Latin Christendom within Eurocentric modernity, and this must not be taken to reduce it or to be speaking of all forms of "Christianity."

upon a communicative or representational model as an assimilative one," where a "labour of love" was sought to shape one's self in that image (Mahmood, 2009: 847). An assimilative model, in this sense, is one where the Muslim moves toward becoming one with the prophet who, in turn, had sought to become one with the entirety of existence and all that is within it. The question, in this "struggle," is accordingly an ethical one where ethics are understood as "the careful scrutiny one applies to one's daily actions in order to shape oneself to live in accordance with a particular model of behavior" (Mahmood, 2012: 187). It was specifically this that centered a synchronization with the universe and nature in all its life and that participants pursued and found impossible under the current urban modernity. It was specifically this "ethical" model that could form the site where lived experiences and nature could be brought back together, where the structures and conditions making the climate catastrophe inevitable become obsolete. This here begins with the self, but in a way where fashioning one's self along particular cosmological and ethical frames is the grounding and basis of systemic transformation.

Participants offered a specific—surely not mainstream—interpretation of Islam that, I would argue, delinks from the modern anthropocentric episteme. This understanding is mostly rooted in specific interpretations of Shia mystic thought or what is known as *Irfan*. The extent to which these understandings and positions extend or cut across Shia and non-Shia Muslim communities within Lebanon but also across the globe remains unclear. My own personal experience as well as research among these communities indicates that it is marginal. The argument is, accordingly, not that these understandings characterize "the Islamic," nor that they are inherent or necessary within it. The argument is, rather, that there was this *possibility* within/through the Islamic and that this possibility existed and was living—even if at the religion's margins.

Indeed, there are various other Muslim traditions and philosophical currents that generate similar, parallel, or converging cosmologies. This particular thinking is not unique nor is it exceptional. Across the globe, Muslim communities have long developed complex cosmologies far removed from Eurocentric modernity and its anthropocentrism—from the *Mawlawiya* traditions of Turkey to Egyptian Sufi circles and to various East and Southeast Asian Muslim interpretations and synthesis with numerous Indigenous beliefs within their respective regions. Many Muslim activists across the world have also developed and presented alternatives to western "development," "sustainability," and man-nature binaries from within Muslim and spiritual positions. Further, these multiple margins, I posit, echo mystic currents across the world's various "world religions." Alongside this dwell many cosmologies and spiritualities that hold similarly generative potential across the globe—including *Samak Kasay* among the Quechua, *Mino Bimawaadiziwin* among the Anishinabe, or *Ubuntu/Ukama* among many peoples of the African continent. While such cosmologies have mostly been silenced over the past 500 years, they survive—increasingly being reconstituted and mobilized for alternative futures.

It is important here to note that there is no significant environmentalist movement in Lebanon, certainly no significant Muslim environmentalist movement. The discussion presented in this paper remains mostly in the space of limited and personal reflections and conversations. It has not materialized into any larger formation. The reasons for this are surely multifaceted, including Lebanon's complex challenging socioeconomic and political realities. An analysis of this is required. How these positions may come together to form such a movement and how they might form conversations, engagements, allegiances, and fusions with the various echoing beliefs across the world are key questions this chapter raises for future research.

Conclusion

Shia Muslim participants in Lebanon expressed significant alienation and estrangement from nature in urban Beirut, one they experienced as destroying livability. Their critique of this model of being in the world was made from within Muslim traditions where categories from "nature" to the "human" are reconceptualized, and *being* is reframed into a journey toward harmony and synchrony with an already-harmonious universe. Such harmony constituted, and was indispensable, for the good pious Muslim life that participants pursue and where a different relationality to nature and *life* is central. In examining this critique and its underlying episteme, a significant disruption of modernist anthropocentrism emerges.

Based on this, the narrative reducing and abjecting the "religious" and the "Muslim" from the ecological and environmentalist conversation can, or should, be challenged. This chapter argues for the possibility of thinking alongside the Muslim and the "religious" or spiritual to reconstitute non-anthropocentric more liberatory models of knowing and being in the world. As this project's participants show, this rethinking is well-underway within everyday lived experiences in the West Asia region. This argument is not to claim participants' Muslim alternative as an answer to the contemporary global intersecting crises of modernity/coloniality—be this in terms of the ecological dimensions or in its multiple other dimensions and injustices. The argument is rather for thinking alongside such efforts as options, taking them seriously, and generatively engaging with them and contributing to them.

References

Baumann, H. 2017. *Citizen Hariri: Lebanon's Neo-Liberal Reconstruction*, 1st ed. New York: Oxford University Press.

Cherkawi, A. 2020. *The Urban Political Ecology of Green Public Space in Beirut*. M.A. thesis, SOAS University of London, London.

Fawaz, M. 2009. "Neoliberal Urbanity and the Right to the City: A View from Beirut's Periphery." *Development and Change* 40(5): 827–52.

Hamade, K. 2014. "Tobacco Leaf Farming in Lebanon: Why Marginalized Farmers Need a Better Option." In Daniel Buckles, Natacha Lecours, and Wardie Leppan (eds.), *Tobacco Control and Tobacco Farming: Separating Myth from Reality*, 29–60. Anthem Press.

Kardahji, N. 2015. A Deal with the Devil: The Political Economy of Lebanon, 1943–75. UC Berkeley. https://escholarship.org/uc/item/4762t40q

Khodr, H., and N. Awwad. 2021. "Politics, Economics, Ricardus Haber and Environmental Policy in Lebanon." *International Journal of Environmental Studies* 79(5): 799–809.

Krijnen, M., and M. Fawaz. 2010. "Exception as the Rule: High-End Developments in Neoliberal Beirut." *Built Environment* 36: 245–59. https://doi.org/10.2148/benv.36.2.245

Laithy, H., K. Abu-Ismail, and K. Hamdan. 2008. Poverty, Growth and Income Distribution in Lebanon. 13. The International Poverty Centre.

Mahmood, S. 2009. "Religious Reason and Secular Affect: An Incommensurable Divide?" *Critical Inquiry* 35(4): 836–62. https://doi.org/10.1086/599592

Mahmood, S. 2012. *Politics of Piety: The Islamic Revival and the Feminist Subject.* Princeton, N.J.: Princeton University Press.

Makarem, H. 2014. *Actually Existing Neoliberalism: The Reconstruction of Downtown Beirut in Post-Civil War Lebanon.* PhD thesis, London School of Economics and Political Science.

Maldonado-Torres, N. 2008. "Secularism and Religion in the Modern/Colonial World-System: From Secular Postcoloniality to Postsecular Transmodernity." In Mabel Moraña, Enrique D. Dussel, and Carlos A. Jáuregui (eds.), *Coloniality at Large: Latin America and the Postcolonial Debate*, 360–84. Durham, N.C: Duke University Press.

Nibert, D. 2012. "The Fire Next Time: The Coming Cost of Capitalism, Animal Oppression and Environmental Ruin." *Journal of Human Rights and the Environment* 3(1): 141–58.

Rustom. 2012. *The Triumph of Mercy: Philosophy and Scripture in Mullā Ṣadrā/ Mohammed Rustom.* State University of New York Press.

Salibi. K. 1998. *A House of Many Mansions.* London: I.B. Tauris.

Traboulsi, F. 2007. *A History of Modern Lebanon.* London: Pluto Press.

Weis, T. 2013. "The Meat of the Global Food Crisis." *The Journal of Peasant Studies* 40(1): 65–85. https://doi.org/10.1080/03066150.2012.752357

Weiss, E. 2016. "Refusal as act, Refusal as abstention." *Web Cultural Anthropology* 31(3): 351–8.

12

Affective Multispecies Resistance
as Radical Imaginations

Abigail Pérez Aguilera

As discussed by Melissa Nelson (2019), "nonhuman nature offers more intelligent and resilient models" for change and survival. However, as part of the embodiments from the colonial continuum, our (human) notions of cooperation and survival are mediated through a capitalist, colonial, patriarchal, and anthropocentric lens that denigrates the nonhuman. This colonial continuum—the ongoing process of colonization transgressing time and spaces—is inserted within institutions, bodies, and imaginations. In this chapter I present perspectives beyond anthropocentrism as possible solutions to this colonial continuum, and emphasize how we can create a framework to understand and exercise affective multispecies resistance as radical imaginations.

Is it possible to build affective relations with nonhumans and the more-than-human world? Should we renounce and neglect our own constitution as humans and "learn from others"? In her work on *ruminant epistemologies*, Lucrecia Masson (2017) does indeed argue that it is possible and necessary to learn from ruminants such as cows. She especially points out that we can learn from their cooperative behavior, their willingness to find a community, as well as the representation of their bodies as anti-capitalist: slow, cooperative, and rebellious. Following this strand of research, in this chapter I rely on affect theory as a conduit to analyze the interactions between human and nonhuman lifeforms as well as the affective relations that may be derived from their constant contact in different times and geographies. We can understand affect and affectivity in the context of these discussions as the feelings and emotions we place onto others and the meaning we grant to emotions such as pain, happiness, sorrow, and grief. The relevance of studying affectivity in relation to environmental disasters and conflict lies in the complexity of reactions that these crises provoke in us. Hence, how we think and feel about *others* expresses the different meanings we give to different persons and defines our interactions with them. In this chapter, I apply affectivity to multispecies interactions in the light of environmental crises, and study how affectivity is placed, imposed, and denied in different times and geographies. I question if it is possible to think of multispecies resistance through affectivity outside the sphere of patriarchy, colonialism, capitalism, and other forms of oppressions. To do so, I rely on discussions around affectivity in relation to compassion, pain, and suffering during environmental conflicts and crises while extending this affect to nonhuman persons and the more-than-human world (Ahmed, 2014).

This chapter first creates a dialogue between affect theory and multispecies relations in the context of the colonial continuum. I particularly focus on how nonhuman interactions have been dealt with inside and outside of academic circles, and reflect on speculations over how our human world is going to exist alongside other species in the wake of the pending environmental catastrophe and environmental activism. Second, I discuss the importance of overcoming human-centric activism and thought through an anticolonial and anti-heteronormative perspective of personhood and under what conditions it is possible to do so in order to recognize and achieve a multispecies perspective.

Affect beyond the Human

The ways we approach multispecies resistance must be based on a critique of the granted notion of the concept and identity of the *human*. In some of my previous work (Pérez Aguilera, 2016a, 2016b), I have discussed how the politics of the intangible and the more-than-human world (Abram, 2012) play a role in how we see, analyze, and live environmental conflicts as well as the multiple forms of resistance that might come from these interactions.

A discussion of multispecies resistance focuses on the multiple entangled worlds beyond the human world. As discussed by Van Dooren, Kirksey, and Münster (2016: 3) "a multispecies approach focuses on the multitudes of lively agents that bring one another into being through entangled relations."[1] For

---

[1] While this multispecies approach embraces an interdisciplinary approach to these complex interactions, it also acknowledges that not all human experiences are the same.

instance, Chao (2021) explains how West Papuans who protested against the Indonesian government appropriated the figure of the monkey, once used as a racial slur against Papuans. Its adaptation to the activists' cause changed the ontological meaning of the monkey. It showed "how animals as political symbols can be creatively reappropriated, repurposed, and redeemed in unexpected ways by subaltern communities as part of their emergent cosmopolitical world-making practices and imaginaries" (Chao, 2021: 282). Multispecies justice is central to multispecies resistance. As other perspectives that rely on non-Western knowledges—in the form of Indigenous and Afrocentric perspectives or knowledge—it is born in the struggle against colonialism, capitalism, and heterosexism, among other forms of oppression (Chao et al., 2022).

The disconnection between the human and the more-than-human world has created abject relations among humans, places, geographies, and other entities that are marginalized to the side of abject nature which modern civilization repulses. By abject nature, I refer to the process by which the concept of *nature* has been created by modern civilization through a series of processes, including the separation of the *natural world* from the human experience. By creating nature as an entity separate from the human world, the human world itself is created. As Silvia Federici (2004) points out in *Caliban and the Witch*, the connection between dispossessing the European peasantry from their land and positing women apart from communal life was the beginning of feudal societies based on the alienation of labor from the land, ending communal life. This resulted in perceiving manual labor (especially that of the peasantry) as dirty, cheap, uncivilized, and uneducated. Communal life (the basis of peasant life before its alienation through feudalism and capitalism) was the way of life for many communities, which encouraged a different relation to the "natural world," including gender relations that allowed women to participate differently within their communities (Federici, 2004).

These mediated relationships to the "natural world" were part of the colonial and settler colonial process which entailed breaking the relationship between Indigenous peoples and their traditional ways of life. This included their relation to the land, the more-than-human world, and their spiritual world, as well as gender relations. All of what Indigenous civilizations consisted of, from the material to the spiritual and the intangible, was intended to be replaced and destroyed through the process of colonization. As pointed out by María Lugones (2010: 745), "the civilizing transformation justified the colonization of memory, and thus of people's senses of self, of intersubjective relation, of their relation to the spirit world, to land, to the very fabric of their conception of reality, identity, and social, ecological, and cosmological organization." For example, as described by Mayan Kaqchikel professor Emma Chirix (2009, 2013), the civilizing transformation in the case of Catholic schools for Indigenous Mayan women in Guatemala had the purpose of eliminating Indigenous ways of life through violent conquest which extended to corporeal politics. This included the imposition of a colonial and Christian form of hygiene that portrayed the Indigenous body as dirty, in need of constant cleaning, and in need of a (Western) civilizing project. As noted by Nancy Perez (2022: 17), "the settler logic and practice of elimination are carried into distinct racial scripts cultivating 'structurally marginalized and racialized workforces' meant to control and disappear marginalized groups from full inclusion in public life." This control and disappearance extend to the more-than-human world, including multispecies relations, where the "public life" resumes to human-mediated nature-cultures.

As Emma Perez (1999: 123) points out when referring to the decolonial imaginary, "to decolonize our history and our historical imaginations, we must uncover the voices from the past that honor multiple experiences, instead of falling prey to that which is easy—allowing the white colonial heteronormative gaze to reconstruct and interpret our past." Through the process of recognizing that some voices and the existence of other worlds have been silenced and denied, we can open a space for the possibility of discussing how it is possible to experience affect toward the more-than-human world in a way that is not grounded in anthropocentric and Eurocentric perspectives. Further, we can see how *affect* necessitates a critique of the violent histories of settler colonialism and the genocidal tendencies of the colonial enterprises. Hence, affect as an emotion that emanates from the human body should be experienced together with a sentiment of responsibility and a sense of affect as praxis.

Why Affect in Multispecies Relationships?

One of the dominant and principal interests around the concept of the Anthropocene and climate change is to explore how we as humans are going to be able to sustain human life during environmental catastrophe. This question gets more complicated once we recognize that this life that many are trying to sustain is part of the "imperial mode of living" (Brand and Wissen, 2021). This concept highlights that living under global imperial capitalism is wasteful and incompatible with goals for a sustainable life. As noted by Brand and Wissen (2021), "the concept of the IML [imperial mode of living] highlights the fact that capitalism both implies uneven development in time and space as well as a constant and accelerating universalization of a Western production model." It further points out that the false solutions provided by capitalism serve as a cover and purposeful delay of the inevitable systemic changes that need to happen in order to stop the climate crisis we are facing.

Following the work of Sylvia Wynter (2003), the modern figure of the "human" is based on Eurocentric, racialized, ableist, and gendered conceptions. As noted by Wynter, the figure of the European man became the measure of the human. If the mere conception of the primordial unit of analysis has been tainted by this colonial conceptualization, how can we have anticolonial discussions about human equality, rights, equal access, and affect? If a hegemonic understanding of what "makes us" human follows a colonial past, then what is left? Alternatively, a more accurate question to ask is: what embodied experiences were left behind? If what defines a *human* is based on a Eurocentric and anthropocentric perspective, then many experiences and knowledges were left behind in the process of establishing Western epistemologies and ontologies. As a result, "hegemonic knowledge/science still privileges modern Western onto-epistemologies and has been effectively instrumentalized for state-centered neoliberal, market-driven, and/or technocratic fixes based on reductionist Eurocentric scientisms" (Figueroa Helland et al., 2021: 44). Hence, as noted by Dhillon (2021), most critiques of environmental depredation and capitalism have ignored that at the center of environmental pillaging is (settler) colonialism and the logics of Eurocentrism.

The tale of the interactions between nonhumans and humans—from the visible to the intangible, to the material, the corporeal, and the spiritual—are mediated by the Western logic of our material relations to nature. For example, as with the Anthropocene narrative, climate change and the catastrophes that the multiple environmental crises have brought are commonly understood in terms of "our (human) survival." This understanding has engendered possible solutions which seem to reproduce the same hierarchical relations with other humans and other species. A multispecies approach is in direct contrast to the anthropocentrism and hierarchy of Western logic. As noted by Van Doreen et al., (2016: 4), such a "[multi-species] immersive approach is now also increasingly being applied to forms of liveliness that many, but by no means all, of us would consider to be nonliving: from stones and weather systems to artificial intelligences and chemical species." Hence, the life that sometimes passes as unseen is what concerns multispecies relations, since "many entities, from geologic formations and rivers to glaciers, might themselves be thought to have distinctive ways of life, histories, and patterns of becoming and entanglement, that is, ways of affecting and being affected" (Van Doreen et al., 2016: 4). Therefore, multispecies relations would allow us to experience a new radical imaginary of the life we deem as inert, passive, and easily disposable and exploitable. As noted by Reinert (2016: 97), if we consider that

> extractive resource capitalism is a sort of ontological machine—an engine that continuously remakes the world and its entities as already-given, in ways that facilitate surplus value extraction—then it is all the more vital to question the paradigms that subtend it and produce *not just nonhuman life but also nonlife as domains of control, use, modification, and productive investment.* (my emphasis)

For instance, the rivers and lake systems of what are now known as Mexico City and the Basin of Mexico provide examples of different ways of resistance from life typically deemed as "nonlife" or "passive life." Water bodies within the colonial imaginary are seen as either sources of pollution, or elements of exploitation and modification. Through the examples from Mexico, we see how, in response to this colonial imaginary, these water bodies have mounted constant resistance to being controlled and forced to disappear from the eyes of the inhabitants of Mexico City by being converted to underground sewage systems. I suggest that the history of the desiccation of rivers and lakes in the Basin of Mexico caused multispecies resistance. Specifically, both water (perceived as "nonlife") and people are fighting against water pollution, floods, violent urbanization, and a modernizing mindset.

Before Spanish colonization, the different Indigenous peoples living in the Valley of Mexico depended on the lakes and river systems for water and food supply. The Basin of Mexico is formed by five lakes (Zumpango, Xaltocan, Xochimilco, Chalco, and Texcoco (where Mexico City was founded)) and forty-five rivers (now mostly serving as part of the public sewage system). The desiccation of the lakes and rivers of the Valley of Mexico by Spanish settlers was justified using arguments of bad smells, bad air quality, insalubrity, and fish being contaminated by the rivers and lakes. In contrast, the rivers and lakes in the Valley of Mexico were more than sources of food and water for the inhabitants of the area. Their knowledge about the behavior of the water, the different food sources in the water bodies of the valley (fish, plants, and birds), and the use of *chinampas* (floating gardens used by Indigenous peoples) were also related to cosmovisions and symbolisms of the water bodies (Tortolero Villaseñor 2002; 164–5). As Espinosa Pineda (1994) explains, the inhabitants of the basin in the Valley of Mexico included animals and plants within the symbolism of Indigenous cosmologies. The variety of aquatic birds in the basin were related to sacred entities. For example, the bird called *ehecatototl* was related to the sacred entity of the spirit of the wind, Ehecatl. Many human and nonhuman relationships derived from the rivers of the basin, which were used for agriculture, transportation via canoes, and more.

These relationships changed with the process of colonization and its aftermath. In 1555, a major flood caused the colonial government to initiate a sewage system to avoid future floods. At the beginning of the twentieth century, by relying on the specialized Western knowledge of medical doctors, engineers, and chronists, the government of Mexico further desiccated the Valley of Mexico based on arguments of public health, as well as potential economic opportunity. Specifically, in 1900, the ruling dictator, Porfirio Díaz, who planned to modernize Mexico, continued the sewage system projects in the city. This included changing the course of the rivers and expanding these projects due to the increasing urban sprawl. According to Mendoza Fragoso (2021), the desiccation of *Lago de Texcoco* and the rivers that form the Basin of Mexico is a "disaster caused by development" since it has affected the Indigenous peoples and the lives that are entangled with them.

Where does multispecies resistance happen in the case of the rivers and lakes of the Basin of Mexico? The constant floods and lack of water supply that Mexico City experiences have been a focus of public policy, environmental activism, and studies of water access and Indigenous rights of Mexico City's *pueblos originarios*. While the water bodies that are part of the Basin of Mexico are seen by city planners and the government of Mexico City as nonlife and lacking agency, the constant floods in the city say otherwise. We can see how water bodies are reclaiming their presence by flooding highways where rivers once ran, collapsing the city and causing almost apocalyptic scenarios when heavy rains affect Mexico City: subway stations flood, cars become trapped in underpasses, and people get waste waters in their homes due to the bad planning of the sewage system.

While the "extractive resource capitalism," as an "ontological machine" (Reinert, 2016), creates nonlife and spaces of absence within the colonial imaginary, multispecies resistance manifests in the multiple ways water bodies find pathways to "escape" the urban planning imposed onto them. Also, the use of the rivers as sewage for urban waste comes back with force, disrupting city life and showing the limits of urban planning and modernization. The desiccation of the Lago de Texcoco and other water bodies in Mexico City have also increased the risk of sinking structures and of buildings crumbling down when experiencing earthquakes. Hence, water bodies make their presence "seen" by neglecting the necropolitical project established by development and modernization. In addition, the loss of water bodies in the Valley of Mexico has caused a disruption in the water supply. In the case of Lago de Texcoco, surrounding communities have lost their water sources and have experienced disruptions in their lives. Many communities, environmental activists, and even urban planners are now trying to recover the rivers of Mexico City and are fighting back against mega-projects affecting the reminiscence of the Lago de Texcoco. In this case, while the resistance does not come from a "species" properly, as Reinert (2016) suggests, it is possible to have nonhuman interactions, or agency, that allow a form of resistance beyond the colonial imaginary.

Affect Theory and Radical Imaginations

One of the key components to building a decolonial strategy that includes a multispecies resistance approach is the recognition that challenges to the anthropocentric mastery of nature (Plumwood, 1993) must include the making of imaginations that extend to nonhumans and the intangible world (Pérez Aguilera, 2016a). Writing on endangered nonhumans and cultures of extinction, Ursula Heise (2010: 6) explains that "understanding the ways in which relationships to other species already form part of our self-understanding will be useful in developing the forms of multispecies justice and multispecies cosmopolitanism." These interactions have been intertwined with the presence and existence of nonhumans and the intangible worlds, including those stories that have been silenced or neglected, such as Indigenous cosmologies and Afrocentric spiritualities.

It is not that affect (as an emotion) is not present in environmental conflicts, but we need to reflect on how we see affect and what comes after it. As Nicole Seymour (2018: 240) points out:

> Affect itself is understood as a question of interrelationality and interconnectivity; most affect theorists define affect as something that does not exist separately on its own and that is not interior to an individual. It is instead that which is generated between—between two bodies, between a person and an object, and so forth—or transferred across, in a kind of "contagion."

Looking for Colonial Fractures in a Multispecies World

One of the critical moments for my work was realizing the incompleteness of my own analysis of environmental justice and environmental racism when several of these movements involved active agents from the more-than-human world. The concept of *fractures* introduced by María Lugones (2010) in her important article, "Towards a Decolonial Feminism," is useful and vital when dealing with a critical approach to multispecies resistance. A key contention of coloniality is that there exists a continuum of lived colonial situations, including imposed hierarchies and positions of subjugation. Hence, according to Lugones (2007: 749, 753),

> In thinking of the starting point as coalitional because the *fractured locus* is in common, the histories of resistance at the colonial difference are where we need to dwell, learning about each other.

> The fractured locus includes the hierarchical dichotomy that constitutes the subjectification of the colonized. But the locus is fractured by the resistant presence, the active subjectivity of the colonized against the colonial invasion of self in community from the inhabitation of that self. (my emphasis)

At the end, what we are looking for is a *fractured locus*. As Lugones points out, these fractures are the resistances within and against the *colonial continuum*. A *fractured locus* is understood as a moment where coloniality is being disrupted by a contestation and resistance. It can be interpreted as a negation of the established paradigms of the colonial/modern gender system (Lugones, 2007) and its connections to the paradigms of Western civilizatory practices (Chirix, 2013), including but not limited to forced heterosexuality, the politics of the body attached to Eurocentric practices, and the annihilation of any other forms of being and thinking that are not rooted in hegemonic knowledges. Hence, to create a multispecies radical or decolonial imaginary is to use the necessary epistemological tools to (re)create a series of allegiances to practices, entities, and worlds that might be neglected or obscured.

The fractured locus we are looking for includes those moments, spaces, practices and recognitions of other existences and possibilities of being and experiencing different worlds beyond what colonialism has presented to us. Finding and recognizing this fractured locus is part of a radical imagination, those spaces of openings for resistance. The existence of a fractured locus also shows that resistance has been a constant within settler colonialism, capitalism, and patriarchy. In the case of multispecies resistance, the goal is to avoid the romanticized notion of nature being resilient or finding common human traits in nonhuman species but to understand that there is a pluriverse of worlds out there which, although unknown, exist regardless of our human experience and making of them.

For example, the environmental conflict in Wirukuta (Mexico) is part of multispecies resistance based on non-Western spiritualities and the recognition of a pluriverse. Wirukuta is a sacred site of the Wixarika or Huichol people, an Indigenous group located in the Sierra Madre Occidental in Mexico. Their territory covers several states including Nayarit, San Luis Potosi, Zacatecas, Jalisco, and Durango. The town of Real de Catorce in the state of San Luis Potosi, close to Wirikuta, has been a site of intensive mining operations, which also included a constant attack on Indigenous Wixarika sacred territories. A well-known illustration of Indigenous cosmologies clashing with Western worldviews, the case of Wixarika, gained attention due to the recognition and inclusion of nonhuman personhood and agentiality in the activism by the Indigenous Wixarika people. The case was presented to the Supreme Court of Justice (SCJN) in Mexico as a conflict that was not restricted to environmental law, human rights, or any other conventional environmental justice case. In 2013, the SCJN issued an injunction, halting mining operations on the sacred site of Wirikuta based on a recognition that within the Wixarika sacred territory, nonhumans and everything that inhabits the territory is sacred (Pérez Aguilera, 2016b).

The pluriverse in relation to environmental conflicts is a matter of recognizing the unseen and neglected life of the more-than-human world as well as the lives of humans who have been excluded from the projects of modernity and development. The strategy used by Wixarika activists and allies was one based on the recognition that another world is living on the sacred territory, challenging non-Indigenous ways of seeing nonhumans. For example, the plant of the peyote, which is sacred for the Wixarika, has been exploited by non-Indigenous people who have been collecting it without permission and misusing it for non-ceremonial purposes. The case of multispecies resistance in the sacred place of Wirikuta needed an interlocutor, the Wixarika people, who spoke for the territory, exposing the limits of Western rationality. This case confronts a worldview that neglects life and sees the exploitation of Wirikuta for productive investment in what the Wixarika see as sacred in their territory (Lamberti, 2018). To radicalize our relations, including our movements of resistance, we need to embrace the idea that we as humans cannot "feel for others" or have an affect toward the more-than-human world without recognizing the presence of the pluriverses on their own terms.

Therefore, to find a fractured locus is to pay close attention to ongoing resistance movements (environmental, feminist, or political) and how these are framed outside the boundaries of hegemonic knowledge. Within this framework, environmental conflicts situated as multispecies resistance movements signify indeed the violent confrontation of two worldviews. The first responds to neoliberal and Western civilizatory practices such as developmentalist projects, growth as an ultimate goal, and precarization of lives through a process of capitalist enterprises. The second, as noted by the feminist Global South collective *Miradas Criticas del Territorio desde el Feminismo*, is land defense rooted in a connection of the body and the land, becoming a body-land entanglement. This implies that the communities practice critical geographies of the land, which includes understanding their connections beyond the tangible and visible and the reclamation of human corporealities as part of a communal (human and nonhuman) allegiance.

Thus, the first step would be to identify the fractured locus in environmental conflicts, which will allow us to see differently—to see the multiple worlds (or pluriverses) that sometimes go unnoticed (Escobar, 2018). Placing multispecies resistance as part of our study of affectivity would open questions of how we care for others that are not visible or have remained unnoticed. In order to have an effective multispecies resistance, it is necessary to rely on non-hegemonic cosmologies in order to expand our affect beyond human-centric perspectives of care and resistance. To do so would require purposefully creating fractured loci where ontologies beyond colonial rationales are recognizable and allow a radical imagination beyond anthropocentric worldviews.

References

Abram, D. 2012. *The Spell of the Sensuous: Perception and Language in a More-than-Human World*. New York: Vintage.

Ahmed, S. 2014. *Cultural Politics of Emotion*. Edinburgh: Edinburgh University Press.

Brand, U., and M. Wissen. 2021. *The Imperial Mode of Living: Everyday Life and the Ecological Crisis of Capitalism*. London: Verso Books.

Chao, S. 2021. "We are (not) Monkeys: Contested Cosmopolitical Symbols in West Papua." *American Ethnologist* 48(3): 274–87.

Chao, S., K. Bolender, and E. Kirksey, Eds. 2022. *The Promise of Multispecies Justice*. Durham, NC, and London: Duke University Press.

Chirix García, E. 2009. "Los cuerpos y las mujeres kaqchikeles." *Desacatos* 30: 149–60.

Chirix García, E. 2013. *Cuerpos, poderes y políticas: mujeres mayas en un internado católico*. Guatemala: Ediciones Maya'Na'oj.

Dhillon, J. 2021. "Indigenous Resistance, Planetary Dystopia, and the Politics of Environmental Justice." *Globalizations* 18(6): 898–911.

Van Dooren, T., E. Kirksey, and U. Münster. 2016. "Multispecies Studies Cultivating Arts of Attentiveness." *Environmental Humanities* 8(1): 1–23.

Escobar, A. 2018. *Designs for the Pluriverse: Radical Interdependence, Autonomy, and the Making of Worlds*. Durham, NC: Duke University Press.

Federici, S. 2004. *Caliban and the Witch*. New York: Autonomedia.

Figueroa Helland, L. E., A. P. Aguilera, and F. Mantz. 2021. "Decolonize, ReIndigenize: Planetary Crisis, Biocultural Diversity, Indigenous Resurgence, and Land Rematriation." In McCullagh Suzanne, Luis I Pradanos, Ilaria Tabusso Marcyan, and Catherine Wagner (eds.), *Contesting Extinctions: Decolonial and Regenerative Futures*, 61–150. Lanham: Lexington Books an imprint of The Rowman & Littlefield Publishing Group.

Heise, U. 2010. "Lost Dogs, Last Birds, and Listed Species: Cultures of Extinction." *Configurations* 18: 49–72.

Lamberti, M. J. 2018. "Los conflictos por la minería en territorios indígenas: hacia una comprensión sociológica no sociocéntrica." *Carta Económica Regional* (122): 31–55.

Lugones, M. 2007. "Heterosexualism and the Colonial/Modern Gender System." *Hypatia* 22(1): 186–219.

Lugones, M. 2010. "Toward a Decolonial Feminism." *Hypatia* 25(4): 742–59.

Masson, L. 2017. *Epistemología rumiante*. Valencia: Feminismo Estrías Autogestión.

Mendoza Fragoso, A. 2021. "La huida de la Sirena. Una narrativa del desastre de la desecación y el despojo en los pueblos ribereños al noreste de la Ciudad de México." *Revista de Antropología y Sociología: Virajes* 23(2): 23–58.

Nelson, M. K. 2019. "Wrestling with Fire: Indigenous Women's Resistance and Resurgence." *American Indian Culture and Research Journal* 43(3): 69–84.

Pérez Aguilera, A. 2016a. *Feminist Decolonial Politics of the Intangible, Environmental Movements and the Non-Human in Mexico*. Arizona State University.

Pérez Aguilera, A. 2016b. "Mining and Indigenous Cosmopolitics: The Wirikuta Case." In Mark Anderson and Zelia Bora (eds.), *Ecological Crisis and Cultural Representation in Latin America: Ecocritical Perspectives on Art, Film, and Literature*, 179. Lanham, MA: Lexington Books.

Pérez, E. 1999. *The Decolonial Imaginary: Writing Chicanas into History*. Bloomington: Indiana University Press.

Pérez, N. 2022. "Red Dust: Migration and Labor as Seismic Fractures to the Anthropocene." *Resilience: A Journal of the Environmental Humanities* 9(2): 14–29.

Pineda, G. E. 1994. "*Las aves acuáticas, un medio prehispánico de interpretación del cosmos*." *Ciencias* (034): 17–22.

Plumwood, V. 1993. *Feminism and the Mastery of Nature*. London: Routledge.

Reinert, H. 2016. "About a Stone: Some Notes on Geologic Conviviality." *Environmental Humanities* 8(1): 95–117.

Seymour, N. 2018. *Bad Environmentalism: Irony and Irreverence in the Ecological Age.* Minneapolis: University of Minnesota Press.

Tortolero Villaseñor, A. 2002. "El agua en la cuenca de México." In P. A. García (ed.), *Agua, cultura y sociedad en México.* El Colegio de Michoacán AC.

Wynter, S. 2003. "Unsettling the Coloniality of Being/Power/Truth/Freedom: Towards the Human, After Man, Its Overrepresentation—An Argument." *CR: The New Centennial Review* 3(3): 257–337.

IV

Conclusion

Connecting Struggles
Final Reflections and Key Insights

Felix Mantz and Mariko Frame

While one could find many lessons and connections from the study of the rich cases presented in this book, in this final chapter, the editors reflect on three key insights, observations, and lessons that emerge and cut across the different contributions. These include (1) the stratified consequences of extractivism and the pluriverse of responses, (2) the complex relationship of different movements to the state, and (3) the need for international solidarity in the face of global capitalism.

Stratification and a Pluriverse of Responses

Resistances against extractivism and imperialism, and the perspectives behind them, take a wide variety of forms. Many of the cases presented in this book fall outside of the resistances that have historically arisen from working-class movements associated with the orthodox Left within the Global North. As the cases in this book illustrate, a broad variety of culturally specific perspectives, worldviews, knowledges, cosmologies, and value systems—a pluriverse—challenge a homogenizing, singular logic of development. They contest hegemonic and dominant ideas of what is desirable for human society and human-nature relations. Likewise, the cases illustrate that a great diversity of resistance strategies are employed, with creative on-the-ground contestations tailored to specific circumstances.

Many of the case studies in this book center Indigenous pathways and "alternatives" of nurturing lifeways and political, economic, cultural, social, and spiritual forms of organization that are at least partially autonomous and independent of the circuits of capital and the state. What is clear is that, in many cases, extractivism, imperialism, land dispossession, and environmental degradation faced by Indigenous communities constitute sites of struggle for land bases, territories, cultures, and entire ways of life. The examples of resistance collected in this volume suggest that such struggles are fundamentally about the material and often interconnected, cultural survival of communities. Struggles against extractivism, land grabbing, colonial conservation, hydroelectric dams, deforestation, and beyond are not merely about the viable livability of a future planet but about questions of the existential reproduction of these societies and their ways of life in the present and future. This is at times very different from climate and environmental movements based in the Global North whose concerns are often not connected to questions of imminent death, loss of livelihood, displacement, and cultural erasure. Recognizing and centering these realities of resistance against extractivism and imperialism in the Global South—which includes the "Souths in the North" such as Indigenous (Chapter 8) and Black-American (Chapter 10) struggles in Turtle Island—is thus key to building genuine and powerful alliances.

For many of these communities, the impetus and aims of their movements appear to center on revitalizing specific ways of life as well as different forms of autonomy from capital, the state, and dominant development paths. As Sherpa's chapter on contesting hydroelectric projects in the Himalayas (Chapter 2) illustrates, Indigenous Himalayan cosmology and Buddhist spirituality are deeply entangled with understandings of appropriate interactions and relations between humans and nature. As such, Himalayan cosmology and Buddhist spirituality become important sources for organizing and nurturing resistance, a process that is paralleled by other cases in the book. Further, in seeking to protect their land and livelihoods, many of the movements illustrated throughout this book pursue high degrees of autonomy and sovereignty from the state and capital. Raghu (Chapter 1) explains how the Pathalgadi Movement sought autonomy by challenging not only extractivism but also state sovereignty and Indigenous politics captured by dominant state, capital, and civil society processes. Similarly, as the representatives of the four (Afro)-Indigenous projects articulated in Chapter 8, what is sought is not better terms from capital or the state but land-based communal self-determination that revitalizes Indigenous knowledge and governance in active opposition to state and capital. However, not all struggles against extractivism presented in this volume seek, or are in a position to seek, territorial autonomy. For example, for the urban poor, perhaps due to strong integration into the circuits of capital, specific political contexts, and histories of material and epistemic colonization, grassroots resistance may be aimed at securing concessions and protections from governments or international institutions.

This pluriverse of responses and horizons is entangled with the multitude of unequal consequences of extractivism and imperialism. Who benefits from these processes, and who loses out, is highly stratified. Hierarchies based on gender, class, race, rurality versus urbanity, and more shape both the nature of outcomes as well as responses to extractivism. While it can be difficult to extrapolate general trends from the small number of cases presented in this book, it is clear that in the face of extractivism, people use the tools they have on hand, in the context of being poor, female, Indigenous, Black, urban or rural-based, and so on. For example, with the tight linkage between elites—political and corporate—found in virtually all cases throughout the Global South, class and, in some cases, caste are major factors in the processes of extractivism and imperialism. As Chapters 6 and 7 illustrate, in Cambodia much of the extractivism and imperialism is directly driven by elites with close governmental ties in the name of development. The poor in every society, whether urban or rural, are more likely to have their rights ignored or be repressed by the state.

Gender and race are central modes of extractivism's social stratification in ways that are irreducible but intimately connected to other systems of power. As Brewer (Chapter 10) and Frame (Chapter 6) point out, women are likely to suffer the most in any given population. They are often poorer, lack land tenure, and are overrepresented in the informal sector and in low-paid, risky, and casual work environments (Chapter 6, Chapter 9, and Chapter 10). Women around the world do higher levels of caring and reproductive labor, housework, and smallholder farming. In fact, women are the world's primary food producers. All of these factors, among others, mean that women are especially impacted by socio-ecological crises and extractive projects. Nonetheless, as these chapters demonstrate, not only are women active participants, they are often leaders in environmental and land-based movements. As Chapter 6 discusses, Cambodian women have successfully led marches, delivered petitions to the government, and protested and resisted land grabbing through tactics such as riots, roadblocks, trespassing, and damaging company property. Through these forms of disobedience, they have provoked international condemnation of land grabbing, and in a few specific circumstances, they have potentially even stymied dispossession of their land and homes. However, it is clear that such resistance is met with severe political and social backlash, including imprisonment, harassment, and even violence from government and private security. The social, political, and economic outcomes of unsuccessful resistance can mean loss of livelihoods and wealth, worsening familial relations, adverse psychological and health outcomes, and more.

From Africa's neocolonial position as exporter of resources to "imperialism . . . [as] simultaneously expressed ideologically in the cultural commodification of Blackness—stereotyped and terrorized in the international division of labor as incapacity, belonging at the bottom of the global hierarchy of nations," Brewer (Chapter 10) and Grant (Chapter 9) identify race at the forefront of extractivism and imperial capitalism. As Grant, following Africana critical theory, argues, the processes of liberation praxis must link revolutionary re-Africanization (resurgence of anti-imperialist culture) and revolutionary decolonization (global linkages among anti-imperialist struggles). For both authors, African peoples and the African diaspora, in particular women, Indigenous peoples, peasants, and urban slum survivors, are primary agents of change and liberation against capitalism, extractivism, and ecocide.

From Engagement to Resistance: Relating to the State

Together, the contributions featured in this volume challenge conventional understandings of political struggle and change that center on the need to capture state power as a pathway for empowerment, rights, recognition, and emancipation. Undoubtedly, in most struggles, the state is engaged with—that is, movements cannot afford to ignore the state, want to dialogue with it, are required to contend with it to some other extent. However, in many cases, movements struggle *against* and *in spite of* rather than *for* or *over* the state.[1] This is no accident. The contributions teach us that wherever we look, capitalism, ecological imperialism, coloniality, and extractivism, are deeply entangled with the state. Put differently, struggles against extractivism in the form of land grabbing, hydroelectric projects (HEPs), deforestation, mining, and fortress conservation often find themselves face to face with the state's monopoly of violence.

For instance, communities fighting for livelihoods, identity, and access to the forest in the Kranh-Bassa Proposed Protected Area in southeastern Liberia are confronted with park rangers who inflict violence, block access to the forest, intimidate community members, and were alleged to be involved in a hunting cartel (Chapter 5). Similarly, land grabbing for conservation schemes in northern Tanzania requires the mobilization of various technologies of state violence against Maasai pastoralists, including borders, prisons, and police (Chapter 3).

---

[1] It is crucial to note that in many of these cases, the state presents itself not as a self-contained actor but rather as an "open" entity which is intimately connected to transnational histories and structures of capital, knowledge, race, and gender.

Direct confrontation with the state extends to the more-than-human as well. For instance, the Mexican state and its desire for modern developmentalism (among other factors) have resulted in the desecration of water bodies and rivers in the Basin of Mexico. This has catalyzed vistas for multispecies resistance—water refuses to be tamed, flooding urban centers, slowly sinking Mexico City from below, and disrupting the daily lives of urban residents (Chapter 12).

The visions for different socio-ecological futures pursued by many movements, projects, and communities featured in this book do not center the full participation in state apparati. Their envisioned futures are often of territorial, ecological, spiritual and cultural self-determination, autonomous and alternative development paths, and noncolonial relationalities with ecosystems, waterscapes, and land bases. For instance, the existence and potential recovery of lived Shia Muslim cosmologies that are incommensurable with the urban estrangement from ecologies in Beirut looks toward the spiritual, not the state, as a path toward socio-ecological harmony (Chapter 11). Even more explicitly, the processes for territories of life presented in Chapter 8 similarly envision the revitalization of (Afro)Indigenous relationalities, cosmologies, and forms of economic and political organizations that work without the state. In the Himalayan state of Sikkim (India), the decentralized water governance systems that are based on Indigenous and Buddhist relations to water and rivers are in direct contrast to the Indian state's vision of modern hydropowered development (Chapter 2). Similarly, after assessing Indigenous, peasant, and slum dweller struggles across Africa, Samuel Grant finds that to oppose ecological imperialism, what is needed is a "commons-based struggle among organizing populations not-aligned with state or capital, but aligned with each other around common objectives" (Chapter 9). In many other cases presented in the book, related visions of autonomy and self-determination are proposed.

Despite these various visions of autonomy, some forms of representation and inclusion might be pursued as well. That is, at the same time as forms of autonomy are advanced and the state is resisted, struggles might seek some type of participation in or recognition by state institutions. This leads to complex relationships with the state. For instance, self-determination is a central objective of the Maasai struggle against land grabbing in Tanzania. Nonetheless, while strategies of resistance such as participatory video aim at realizing this self-determination and the building of Pan-African unity across Indigenous struggles, they also seek to force a change in the government's attitude toward the Maasai (Chapter 4). Similarly, on the one hand, resistance against the deforestation of the Prey Lang rainforest in Cambodia clearly struggles against the state and corrupt government officials. As the Cambodian activist Narith Nou explains: "We need to change the government system in Cambodia" (Chapter 7). Nonetheless, the Prey Lang Community Network (PLCN) also seeks to pressure the Ministry of the Environment to make the Prey Lang forest into a protected area. Further, the aim is not necessarily self-determination or autonomy from the state (Chapter 7). Similarly, the women activists of the Boeung Kak Lake eviction both protested for changes in national legislation as well as submitted complaints to the World Bank (Chapter 6).

In the case of community organizing against HEPs in Sikkim, the Indian state is driving HEPs and is clearly resisted while some forms of self-determination are pursued. However, strategies and tactics of resistance frequently include petitioning, pleading with, or otherwise speaking to the state (Chapter 2). Finally, the Pathalgadi Movement against state and corporate mineral extractivism in Jharkhand (India) set up autonomous infrastructures such as self-defense committees, banks, and schools in a refusal to assimilate to state-sanctioned forms of neoliberal politics. Posing a direct challenge to state sovereignty, the movement could not be squared with (neo)liberal forms of politics centered on inclusion and representation (Chapter 1). Raghu (Chapter 1) argues that while the Movement's political strategies featured engagement with liberal democratic institutions, it also clearly exceeded them, constituting a non-/anti-liberal politics. In many ways, this understanding of socio-ecological struggles and their engagement with the state can be applied to other cases examined in the book as well. In sum, an engagement with the state should not be read as the adoption of a liberal politics of inclusion but a much more messy, unruly, and complex process that is part of a "repertoire of subaltern ungovernability" (Raghu, Chapter 1).

The Need for International Solidarity in the Face of Global Capitalism

The cases in this book represent a mixture of successes and failures: they teach us of both successful alternatives and resistances but also highlight that as long as the dynamics of global capitalism, coloniality, and extractivism exist, communities (human and nonhuman) will remain under threat. While responses are all contextualized, involving specific places, histories, ecologies, and peoples, many communities face issues that are global in scale. They are connected by a global order that features capitalism and its systems of trade, finance and investment, coloniality, Eurocentric ideologies of progress and development, and racial and gendered hierarchies. How this order and its deep structures manifest varies across different spaces and contexts. As we can see from the contributions in this book, socio-ecological crises are never "locally" contained. They are place-based and context-specific, but both the dimensions of extractivism as well as the modes of resistance transcend scales and involve the global relations of capitalism. For instance, extractivism in Jharkhand (India) is characterized by an entanglement of domestic Indian capital and transnational ecological imperialism. Specifically, Indian multinational conglomerates that drive mineral extraction, such as the Adani Group, are part of a global capitalist order—their strategies for capital accumulation are global, and they are involved in extractivism beyond India (Chapter 1). Similarly, land grabbing for conservation regimes in Tanzania is supported materially and ideologically by Northern and international conservation agencies as well as a global tourist industry (Chapter 3 and Chapter 4).

Resistance against extractivism also frequently takes on global dimensions while responding to place-specific socio-ecological crises. For example, the Garifuna struggle against offshore oil drilling in Honduras, which is led by women, actively builds international solidarity. Garifuna organizations have established relations with activists from the African diaspora in the United States, resulting in (among others) US activists traveling to and joining the Garifuna struggle in Honduras (Chapter 10). Similarly, efforts to build biocultural territories of life across Turtle Island and Abya Yala have resulted in the building of alliances, networks, and transcontinental intercambios de saberes (wisdom/knowledge exchanges) (Chapter 8). Such efforts are paralleled on the other side of the Atlantic. Participatory Video (PV) collectives in East Africa that advocate for Indigenous people's self-determination are actively building Pan-African alliances to learn from one another, share knowledge, and build strong transnational networks. Through the tool of PV, these collectives also seek to engage with transnational audiences, and Northern-based organizations and peoples. For instance, Maasai engagements with PV have resulted in projects and collaborations in the UK to decolonize cultural spaces there (Chapter 4). Similarly, various organizations scoped in Chapter 9—from the Rural Women's Assembly to the Indigenous Peoples of Africa Coordinating Committee—also organize for land, women's, and Indigenous rights across the African continent.

Because of the global dimensions and trans-scalar nature of socio-ecological crises, what is required to put an end to this anthropocentric, colonial, capitalist, racist, and patriarchal order and the extractivism it requires to sustain itself, is transnational alliances, organizing, and movements. This is precisely what Rose Bewer emphasizes when she calls for "deep relationality" and "international solidarity" (Chapter 10). As these cases teach us, winning one specific place-based struggle will likely not be sufficient to end this order because communities will constantly be under threat from the state, global capital, and the transnational manifestations of race, gender, colonialism, and anthropocentrism. Rather, winning arguably requires the success of many movements in different places, through a diversity of tactics and through some level of transnational coordination. That being said, the studies presented here also illustrate how winning specific place-based struggles are indispensable and of the highest importance, not least because they can dramatically improve the lived realities of the people(s) involved. As the various chapters highlight, most forms of resistance emerge from the immediate need for survival, dignity, land, livelihoods, intact ecosystems, and cultural and spiritual revitalization. International solidarity, organizing, and movement-building must be critically attuned to (and directly advance) these goals. The intercambio de saberes between the different processes from Abya Yala, which make up Chapter 8, is a beautiful illustration of this mode of organizing. While the spokespeople of the respective struggles weave their narratives together in the second part of the chapter, they also insist that their projects cannot be subsumed to one another. Hence, the first part of the chapter presents their respective struggles for territories of life in their own voices, separately and on their own terms.

Orthodox Left politics of the past have had a tendency toward homogenization and advancing (inter)national unity by flattening the needs, goals, and aspirations of highly diverse peoples and subsumming them to a singular logic of revolution. The chapters in this volume suggest that in the context of resisting extractivism, this is impossible: it would likely come at the unacceptable expense of erasing the desires, aspirations, knowledges, histories, and efforts of a multitude of peoples. This would constitute a continuation of colonial histories of erasure which many of them have been subjected to for the past 500 years. Hence, when considered in dialogue and relation with one another, the chapters point to the need for global resistances against extractivism and the current world order that reject a singular logic. Pursuing one singular development path, one singular future, and one singular world is incommensurable with the different demands, desires, and visions pursued by highly diverse grassroots movements. It seems that a more appropriate and just framework for global solidarity has to be radically pluriversal—creating a world in which many worlds fit.

Finally, reflecting on the contributions to this volume, we note several limitations, two of which seem especially significant. First, the volume does not feature socio-ecological struggles that center on movements or projects which organize themselves as workers. Workers movements are, however, an important constituent and force in resisting extractivism and socio-ecological crises. The Movimento dos Trabalhadores Rurais Sem Terra (MST), the Brazilian Landless Workers Movement, is perhaps the most prominent of such movements. Second, this book is missing resistances from the Pacific, including East Asia and Oceania. This is by no means a reflection of the absence of extractivism and resistance to extractivism in these geographies. While this book is but a collection of specific snapshots of struggles across the globe, we hope it nonetheless offers insight and lessons for community organizers, scholar-activists, and people of conscience to draw on and consider as we agitate and organize for more socially just and ecologically balanced worlds.

Contributors

Abigail Pérez Aguilera is an assistant professor and interim chair of the graduate program on environmental policy and sustainability management at The New School in New York City. Her current research is on radical imaginations beyond the Anthropocene, militarization and Indigenous resistance, water ecology/justice, and toxic environments. She is part of the Latin American Observatory for the Environment. She is particularly interested in the intersections of multispecies resistance, militarization, discourses around extinction and protection, and anti-colonial projects.

Rose M. Brewer, PhD, is the Morse-Alumni Distinguished Teaching Professor and past chairperson of the Department of African American & African Studies, University of Minnesota-Twin Cities. She is an affiliate faculty member in the Departments of Sociology and Gender and Women and Sexuality Studies. An activist scholar, Professor Brewer publishes extensively on Black radical feminism, political economy, social movements, radical ecology, race, class, gender, and social change. She was a founding board member of Project South: Institute for the Elimination of Poverty and Genocide, a past board member of United for a Fair Economy, and a founding member of the Black Radical Congress. She is the principal investigator of the Humanities Without Walls study, "Environment Justice Worldmaking." And she is the 2024–2025 President of the Society for the Study of Social Problems (SSSP). As a core organizer of the 2007, 2010, and 2015 US Social Forums, the struggle for social transformation in those Forums centered the environmental justice fights of frontline communities. Indeed, her commitment is to change the world for people and the planet.

Mileida Correa is an Indigenous Zenue and a fish farmer. Mileida has been a member of the Board of Directors of ASPROCIG (Colombia) and has served as treasurer. Her work has included foregrounding the leadership and key role of women in the organizational processes of the communities and grassroots organizations with whom ASPROCIG articulates its processes.

Leonardo E. Figueroa Helland, PhD, is an associate professor of environmental policy and sustainability management at The New School University, New York City. He is an associate director of the Tishman environment and design center, where he leads the Indigeneity and Decolonizing Sustainability program. A decolonizing scholar of mestizo (mixed-blood) heritage (Indigenous Mesoamerican and Euro-American), his work underlines the centrality of Indigenous resurgence and revitalization in addressing planetary crises, achieving climate justice, and materializing systemic change. He does so by articulating radical Indigenous approaches with other counterhegemonic liberatory perspectives to envision and enact decolonial futures against and beyond imperialist, settler colonial, neocolonial, patriarchal, anthropocentric, capitalist, and state-centric orders. His articles appear in academic journals such as the *NYU Environmental Law Journal, Journal of World Systems Research, Perspectives on Global Development and Technology, Bajo el Volcán. Revista del Postgrado en Sociología de la BUAP, Studies in Twentieth & Twenty-First Century Literature, Interventions: International Journal of Postcolonial Studies, Journal of Critical Education and Policy Studies,* and *UNESCO Journal of Higher Education.* He has also co-edited two special issues of the academic journal *Perspectives on Global Development and Technology* on "The Earth Crisis and the Global Environmental Movement" and has authored or co-authored chapters in academic volumes such as *Social Movements and World-System Transformation, Inhabiting the Earth: Anarchist Political Ecology for Landscapes of Emancipation, Contesting Extinctions*; and forthcoming in the *Oxford Handbook of Comparative Historical Sociology,* the *Oxford Handbook on Climate Action,* and *Anti-Colonial Food Systems.* He is currently working on a book project under contract with Routledge publishers, prospectively entitled *Indigenous Resurgences & Earth Crises: Decolonizing Pathways to Regenerative Emancipation.*

Mariko Frame is an associate professor of economics and environmental studies at Merrimack College. She has written widely on ecology, economics, imperialism, and development, including the book *Ecological Imperialism, Development, and the Capitalist World-System: Cases from Africa and Asia.* She has lived, researched, and taught in Asia, Africa, and Latin America. She is currently researching the impact of capitalism and globalization on biocultural diversity, traditional environmental knowledge, and eco-cosmologies in Asia.

Samuel Grant has been an organizer working through the intersections of environmental, economic, racial, gender, and cultural justice for decades. Since

1990, he has been a faculty at Metropolitan State University, United States, where he created the minor in community organizing and development. He is the Executive Director at Rainbow Research and manages a small worker coop focused on deep democracy. He is also a fellow of the Institute of the Environment, University of Minnesota, United States. The focus of most of his work now is re-articulating the climate crisis as the *Apartheidocene* and building on the work described in this book to organize formations advancing climate and environmental justice.

José (Angun) Gualinga is an Indigenous Kichwa leader from Sarayaku. He is a member of the advisory team of the governing council of the Kichwa people of Sarayaku. Currently he is Coordinator of the Plan to Strengthen the Declaration of Kawsak Sacha Living Forest, Agenda 2021–2023. Previously, he was Tayak Apu (President) of the Kichwa Indigenous People of Sarayaku. He is also a representative in Europe of the Organization of Indigenous Peoples of Pastaza, currently PAKIRU, and author of the *Path of Flowers Plan or Frontier of Life*; the *Kawsak Sacha Declaration Proposal—Living Forest*; and the *Kindy Challwa, Canoe of Life*, which currently rests in the Museum of Man in Paris, France.

Ali Kaba is a research fellow at the Ducor Institute and a PhD candidate at the American University. His research interests focus on the political economy of customary land and resource governance, particularly in relation to territoriality, identity, peace, conflict, and sustainability. His dissertation specifically examines youth access to land and resources within the context of authority and representation in customary governance. Ali has authored numerous briefs and articles and co-authored two books.

Dr. Ali Kassem is a lecturer in sociology at the National University of Singapore, Singapore. Ali was previously a postdoctoral research fellow at the Institute for Advanced Studies and the Al-Waleed Centre at the University of Edinburgh, where he also taught at the School of Social and Political Science. Prior to that, he was a postdoctoral research fellow with the Arab Council for the Social Sciences and the Carnegie Corporation of New York affiliated with the Beirut Urban Lab at the American University of Beirut. He obtained his PhD in sociology from the University of Sussex, UK. His book manuscript is titled *Islamophobia and Lebanon: Visibly Muslim Women and Global Coloniality* (2023), and he is currently editing a book titled *The Colonial and Arab-Majority Worlds: Across, In-between, and Beyond* (2025).

Haider A. Khan is currently a distinguished university professor and professor of economics at the Joseph Korbel School of International Studies, University of Denver. He has served as the chief international adviser to Arab trade and human development in Cairo, a senior economic adviser to UNCTAD in Geneva, and a distinguished visiting professor to many Asian, European, and Latin American universities. He has published more than twenty books and over two hundred articles in professional journals and received many international awards. Prof. Khan is also an award-winning poet, translator, and literary, and music and art critic.

Nick Lunch is an experienced participatory video facilitator, working at home in Oxford and in over twenty countries since 1996. His passion for the methodology grew from experimenting with handing over control of the camera and film project to Indigenous youth in Nepal. Nick co-founded InsightShare with his brother Chris and has helped lead the organization through an eventful journey of ongoing learning and evolution. Today, his focus is on honing the organization's capacity-strengthening skills and working with community animators to support a growing grassroots participatory video movement. Much of his fieldwork has involved repeat visits to Indigenous partner communities in Africa, Asia, and South and Central America; working together to preserve and promote traditional knowledge; advancing the cause of Indigenous rights; and supporting local representatives as delegates at key global gatherings. Nick has completed accredited training in Nonviolent Communication, Theory U (Presencing Institute), and Permaculture Design.

Felix Mantz works and researches in the Department of Political Science at the University of Hawai'i at Mānoa. His research is transdisciplinary and examines the myriad ways in which the global political economy, planetary environmental crises, and colonial structures of power are entangled. Drawing on a range of critical theories, especially anti/decolonial, anarchist, and Indigenous thought, Felix is particularly interested in questions of land, autonomy, food systems, decolonization, and extractivism.

Angela Martínez, MS, is the Amazon Defenders Fund Director for Amazon Watch. Angela has over three decades of activism and international field experience accompanying Afro-descendant, LGTBIQ+, women, youth, and Indigenous-led movements in the Abya Yala (Latin America and the Caribbean). She has worked on rights of nature and environmental justice, civil and political rights, and sexual health and rights through mobilizing solidarity, ethnographic and qualitative research as accompaniment, advocacy, and capacity building. Angela's roots trace back to the Hñahñu people of Hidalgo, Mexico, where both her grandmother and mother were born, which drives her conviction to accompany Indigenous-led movements to advance their agenda of self-determination and autonomy. She has an MS in environmental policy and sustainability management from The New School in New York City, which she obtained alongside the John Clinton Award for her trajectory as a student and practitioner and for her MS Capstone research project.

Neftalí Reyes Méndez is a coordinator of territorial rights for the organization Services for an Alternative Education A.C (EDUCA AC, Oaxaca). Neftalí is the son of Zapotec and Mixtec. He is a popular educator, human rights defender, and community advisor in defense of territories in Oaxaca, México. Neftalí has co-directed, coordinated, and contributed as coauthor to various projects and publications such as *Saberes comunitarios: alternativas de vida frente al modelo de desarrollo en Oaxaca* (Community knowledge: alternatives for life in the face of the development model in Oaxaca).

Samwel Nangiria is the founding director of Oltolio Le Maa, a Maasai-led participatory video group that documents human rights issues, social development, and Maasai culture. Samwel began using participatory video to defend Maasai territories from land grabs and forced evictions of Maasai pastoralists, and in 2017, he was acclaimed as Tanzania's Rural Human Rights Defender of the Year. Samwel also uses participatory video as a tool for preserving Maasai culture and redressing colonial narratives of Indigenous peoples in Western or globalist institutions. He spearheaded the Maasai Living Cultures project (a partnership between InsightShare and the Pitt Rivers Museum Oxford), which dismantles colonial narratives within museums by empowering Indigenous peoples to curate their own narratives. He is also the co-founder of the Pan-African Living Cultures Alliance.

Juan José López Negrete is an afro-descendant and Indigenous Zenue, fish farmer, photographer, engineer, and specialist in environmental education development. He coordinates the Association of Farmers, Indigenous, and Fisherfolks of the Ciénega Grande Del Bajo Sinú, ASPROCIG (Colombia).

Narith Nou is a former Prey Lang Community Network (PLCN) Project Coordinator, a grassroots network of local community members working to save the Prey Lang forest. PLCN is the recipient of numerous international prizes, including the International Society for Tropical Foresters Innovation Prize, awarded by the Yale School of Forest and Environmental Studies, and the United Nations Development Program (UNDP) Equator Initiative Prize, among others.

Pratik Raghu is an assistant professor of environmental justice and sustainability at Saint Mary's College of California. Pratik completed his PhD in global studies at the University of California, Santa Barbara, in 2022, focusing his dissertation on semi-autonomous Indigenous mobilizations against neoliberal dispossession and state violence in Jharkhand, India, and Oaxaca, Mexico. He is broadly interested in anti-capitalist, anti-authoritarian, Indigenous, and decolonial grassroots movements and theoretical frameworks across the Global South, with a geographical emphasis on South Asia and Latin America. His writing has appeared in both recognized scholarly journals such as *Perspectives on Global Development and Technology* and on popular public intellectual platforms such as *ROAR Magazine*, *CounterPunch*, and *Society & Space*.

Angelica Castro Rodriguez is a sociologist and researcher with Services for an Alternative Education A.C (EDUCA AC, Oaxaca), where she has a directing role in the areas of public engagement and citizen participation. Angelica has co-directed, coordinated, and contributed as coauthor to various projects and publications such as *Saberes comunitarios: alternativas de vida frente al modelo de desarrollo en Oaxaca* (Community knowledge: alternatives for life in the face of the development model in Oaxaca) and *El diagnóstico sobre agresiones a defensores y defensoras comunitarias y el papel de los organismos públicos de derechos humanos en Oaxaca* (Diagnosis of aggressions against community defenders and the role of public human rights organizations in Oaxaca).

Dawa Yangi Sherpa is a Sherpa-Nepali woman and an advocate for Indigenous peoples' worldviews, biocultural diversity, and environmental justice. She holds a master's degree in environmental policy and a bachelor's degree in biology. She uses a decolonial approach to address critical issues such as ecological debt, water and climate injustice and critiques false solutions to climate crises affecting the Global South. She works as the Social Safeguards Technical Specialist at the Wildlife Conservation Society (WCS), where she advocates for the rights of Indigenous peoples and local communities within global conservation amid the ongoing crises of climate change, pollution, and biodiversity loss. Dawa is a board member of the United Sherpa Association, Inc. (2023–2025) and Greens REALIGN (Redistributing Environmental Authority, Leadership, and Influence to Grassroots). She also serves as a group of advisors member (2024–2027) for the tenure facility, contributing her expertise to promote equitable land and resource rights.

Baba Sillah, a PhD candidate in global studies at Sophia University, Tokyo, Japan, takes an interdisciplinary approach to his research. His work investigates the political-economic and socio-complexities of depeasantization, situating the crisis within the current phenomenon of land grabbing. His research spans agrarian studies, comparative politics, political governance, economic development, political ecology, state-society relations, and global capitalism.

Çaca Yvaire (Atakapa Ishak) is an Afro-Indigenous territorial scholar-practitioner and student of planetary science, political geoecology, and the arts. Therein, he brings a unique lens to conservation work. He is primarily invested in stimulating and uplifting an infrastructure that supports a rich and imaginative solidarity between lifeways and honoring the varied traditions of relating sustainably to the world around us through non-industrial agroecology, agriculture, silviculture, aquaculture, etc. that have been historically sidelined by imperialist design. Çaca applies black, creole, Indigenous, and radical cosmologies in practical cultural "mechanisms" and organizational strategies for re-uniting and re-turning health with fragmented communities—both human and non-human. Whether developing cultural respect easements, evaluating rainmaking ceremonies, supporting grassroots activism, or drafting plans for transition, Çaca is committed to the freedom of present and future BIPOC children to grow, learn, love, laugh, and migrate as (and when) necessary, without the terrors and losses known to BIPOC ancestors. Çaca has also been Terran Shield: Community Conservation Co-director for the Northeast Farmers of Color Land Trust (NEFOC). Under the title of the Terran Shield, he is defending a space in which a regional and national community of land stewarding (water-singing, earth-building, fire-wielding, air-breathing) practitioners of all creeds may collectively prepare and secure a place for independence and resiliency in a time of wicked cultural extraction, neglectful and uneasy political leadership, and well-funded environmental destruction.

Index